T0177220

CONTEXTUALIZING OPENNESS

University of Ottawa **Press**
Les **Presses** de l'Université d'Ottawa

The University of Ottawa Press (UOP) is proud to be the oldest of the francophone university presses in Canada and the only bilingual university publisher in North America. Since 1936, UOP has been "enriching intellectual and cultural discourse" by producing peer-reviewed and award-winning books in the humanities and social sciences, in French or in English.

Library and Archives Canada Cataloguing in Publication
Title: Contextualizing openness : situating open science / edited by Leslie Chan; co-edited by Angela Okune, Becky Hillyer, Denisse Albornoz and Alejandro Posada. Names: Leslie Chan, 1959- editor.
Description: Series statement: Perspectives on open access | Includes bibliographical references.
Identifiers: Canadiana (print) 20190104767 | Canadiana (ebook) 20190104856 |
ISBN 9780776626666 (softcover) | ISBN 9780776626673 (PDF) | ISBN 9780776626680 (EPUB) |
ISBN 9780776626697 (Kindle)
Subjects: LCSH: Communication in science—Developing countries.
Classification: LCC Q225 .C66 2019 | DDC 501/.4—dc23

Legal Deposit: Third Quarter 2019
Library and Archives Canada

Copy editing	Nord Compo
Proofreading	Heather Lang
Typesetting	Nord Compo
Cover design	Édiscript enr.
Cover image	Shutterstock

A co-publication with
International Development Research Centre
PO Box 8500, Ottawa, ON, K1G 3H9, Canada
www.idrc.ca / info@idrc.ca
ISBN 978-1-55250-611-0 (IDRC e-book)

The research presented in this publication was carried out with the aid of a grant from the International Development Research Centre, Ottawa, Canada. The views expressed herein do not necessarily represent those of IDRC or its Board of Governors

The University of Ottawa Press gratefully acknowledges the support extended to its publishing list by Canadian Heritage through the Canada Book Fund, by the Canada Council for the Arts, by the Ontario Arts Council, by the Federation for the Humanities and Social Sciences through the Awards to Scholarly Publications Program, and by the University of Ottawa.

u Ottawa

CONTEXTUALIZING OPENNESS
Situating Open Science

EDITED BY

Leslie Chan, Angela Okune, Rebecca Hillyer,
Denisse Albornoz, and Alejandro Posada

University of Ottawa Press

International Development Research Centre
Ottawa • Amman • Montevideo • Nairobi • New Delhi

2019

Table of Contents

Section 4: Open Science for Social Transformation

Preface

This is the inaugural volume of the series "Perspectives on Open Access." Since 2009, the University of Ottawa has made a strong commitment to supporting and promoting Open Access initiatives, and the University of Ottawa Press has developed a significant collection of titles that are openly available. The University of Ottawa Library has been a key partner in providing financial support for selected new books to be openly available upon publication, in keeping with the library's commitment to support open dissemination of knowledge.

The goal of this new series is to explore the transformative implications of Open Access philosophy and practice in its economic, social, cultural, and political dimensions. As an emerging and vital area of scholarly inquiry, Open Access is playing a growing role in shaping public policy and the values of contemporary society. This timely book focuses on the role of Open Science in today's world. It brings together the collective learning and knowledge from twelve research projects that formed the Open and Collaborative Science in Development Network (OCSDNet). This network engaged in a wide variety of research and practices in many countries to explore and demonstrate the benefits and limitations of Open Science principles and practices in various Global South contexts.

At its heart, Open Science seeks to bring about a re-evaluation of the role of science in our rapidly changing world. It critiques the status quo of knowledge production by asserting the importance of democratizing knowledge, by reassessing the power relations in our knowledge infrastructure, and by arguing that scientific knowledge needs to be managed in collaboration with those who help generate it and will benefit from it. As such, it raises questions about the role of governance in scientific knowledge infrastructure, the need for a re-evaluation of the research agendas that drive institutional and societal priorities, and the urgent implications of the digital information divide between the North and the South. It builds upon the insights of the Open Access movement on knowledge production and extends this in new and important scientific and social directions.

This global-scale volume captures the experience and outcomes from research projects in Lebanon, Kenya, Haiti, Brazil, Argentina, Kyrgyzstan, Southwest Asia, and elsewhere. It covers a very broad range of issues—water quality testing, disaster recovery planning, a biodiversity databank, Indigenous people's knowledge, intellectual property rights, environmental education, citizen science, and sustainable local development. While each of these projects is specific in its goals and circumstances, they all share the values of a new paradigm of science that is open, collaborative, and inclusive. As the chief editor, Leslie Chan, writes, "The ability to participate, to connect, and to co-produce knowledge with others who share common concerns is far more important than simply access to content or resources." The researchers behind these projects also share a belief in the critical importance of applying Open Science thinking to developing sustainable solutions for environmental, health-related, and socioeconomic issues that affect people everywhere.

As such, this framework of openness in science embodies a keenly ethical dimension. It raises pointed questions of social justice and the legitimacy of scientific purpose and action while incorporating diverse forms of knowing and knowledge distribution into scientific practice. It highlights the opportunities that openness can achieve while remaining very sober about the challenges that need to be overcome. It is a seminal work that will contribute significantly to the global conversation on the role of science and knowledge in our world. We are very proud to publish it, and we hope that you will find it engaging, provocative, and inspiring.

Tony Horava, Series Editor, Perspectives on Open Access

INTRODUCTION

CHAPTER 1

Situating Openness: Whose Open Science?

Leslie Chan

Open Science is the idea that knowledge from across different domains should be openly shared as early as it is practical in the research process (Nielsen 2011). Extending beyond the discourse on Open Access, which has focused on free online access to research outputs (Chan and Gray 2014), Open Science proposes to expand access to and participation in the processes and outputs of the entire research life cycle (Bartling and Friesike 2014; Friesike et al. 2015). This also implies that an expanded range of actors, including "citizens," could take part in the knowledge production process, from agenda setting to research design, and from the dissemination and uptake of research to subsequent policy influence (Chan et al. 2015).

When placed in the global context, this view of Open Science inevitably leads to important epistemic questions about the nature of science and knowledge: Whose science is being open? By whom? Who is going to benefit from these new framings and practices? What are the risks? Will this lead to equality and equity of knowledge access and production by researchers in unequal settings? Will Open Science disrupt the existing global power structure of knowledge legitimation? Will it lead to further marginalization of knowledge from the Global South? How will Open Science contribute toward the Sustainable Development Goals?

These questions push us to confront more fundamental questions of what constitutes scientific knowledge, and how to reframe incentives consistent with the value and culture of knowledge sharing and collaboration that is at the heart of the idea of knowledge commons, the model in which individuals and communities have access to the mechanisms and autonomy that enable them to decide how their collective knowledge will be used, shared, governed, and cared for (Hess and Ostrom 2006; Bollier and Helfrich 2014; Frischmann et al. 2014).

The attempt to answer these questions was one of the key motivations underlying the formation of the Open and Collaborative Science in Development Network (OCSDNet) in 2014, which had the primary aim of understanding what Open Science means, and the conditions under which its principles and practices could contribute to development thinking, practices, and positive changes in local and translocal development outcomes.

This volume brings together the collective learning and observations by the twelve research projects that formed the OCSDNet, which for two years (2015–2017) engaged in research and participatory activities to understand the benefits, potential, and limitations of Open Science principles and practices in various Global South[1] contexts. The primary aim of this collection is to present case studies, empirical observations, diverse conceptual perspectives, and critical reflections on how opportunities and challenges posed by Open Science vary across geopolitical regions. This further allows us to identify key differences and similarities across institutions, infrastructure, and governance of knowledge production and knowledge-based resources in diverse settings in the Global South.

OCSDNet: Structure and Methodologies

The OCSDNet is a collective of diverse research endeavours that were brought together under the project titled "Catalyzing Open and Collaborative Science to Address Development Challenges," funded by Canada's International Development Research Centre (IDRC) and UK's Department for International Development. A team consisting of representatives from iHub in Nairobi and researchers from the Centre for Critical Development Studies at the University of Toronto Scarborough assumed the management of the network and sub-grantees, as well as research coordination and analysis of the data collected from the participating projects.

From its onset, the network was also supported by a team of international expert advisors[2] who each served as mentor for a set of projects. The advisors also assisted the coordination team in re-fining the initial research framework, evaluating proposals from the sub-grantees, and providing strategic advice on how best to utilize the rich observations and resources within the network to advance the research objectives. The advisors had been closely involved with all the face-to-face network meetings as well as the production of this volume through reviewing the chapters and providing introductory remarks for the various sections.

The OCSDNet sub-projects were selected through a broad open call for concept notes in July 2014.[3] We received over ninety submissions on a broad range of initiatives from around the world and selected fifteen projects to take part in a proposal development workshop in Nairobi in October 2014. A two-month online interac-tive proposal development phase followed the workshop, and final approval of the twelve projects was made in early 2015, with each project receiving funding for a two-year period.

Three of the twelve projects were based in Sub-Saharan Africa, one was from the Middle East, one from the Caribbean, four from Latin America, and three from South, East, and Central Asia.[4] Together, researchers were distributed across twenty-six countries. The teams were composed of individuals with highly diverse academic and practical backgrounds, including law, performance art, education, climate change research, science and technology studies, the maker movement, intellectual property rights, biodiversity, health, and en-vironmental conservation.

In addition, the projects were carried out by a broad range of research actors, from young or early career researchers to those with well-established records and deep local and international network-ing experiences; and from adult community participants to school children. The institutional actors were also diverse, from small-scale independent NGOs and loosely organized grassroot communities to formal research organizations with established international partners. The variety of geographic, institutional, and subject areas provided rich opportunities for case studies as well as for comparative analysis. Importantly, the diversity of research collaborators and participants deeply enriched the findings and the conceptual perspectives pre-sented in this volume.

Over the course of two years, using an array of research meth-ods, each project team explored the challenges and opportunities for imagining science as open and collaborative as well as the potential of Open Science to contribute toward inclusive and sustainable development in their local contexts.[5]

Research Questions and Objectives

The overarching research question for the network was whether, and under what conditions, open and collaborative science practices could lead to development outcomes and community well-being. The longer-term goal was to contribute to the building of a new field of study (Open and Collaborative Science or OCS), stimulate production of evidence to inform policy and practice, and nurture a community of researchers who identify themselves as working on OCS for development. While the researchers from the sub-projects served as collaborators of the network, each project also served as a case study by providing empirical observations and reflective learning for the network synthesis and the overall understanding of what openness means across the various knowledge production contexts.

From the inception of the network, we were deliberate not to impose a specific definition of "Open Science," nor did we prescribe what constitutes development outcomes. We encouraged applicants to think broadly about how they would define openness and science, as well as development, according to their local contexts. The primary intention was to use a grounded theory approach (Glaser and Strauss 1967; Charmaz and Belgrave 2015) to see what common understandings of Open Science would emerge, which would in turn allow us to develop conceptual frameworks to deepen our understanding of "openness"—not only with regard to Open Science, but across the broad spectrum of discourses on openness, such as Open Access, Open Educational Resources, Open Data, and Open Innovation. The secondary goal was to use the results from the grounded method to develop a theory of change, and to better understand how change happens with regard to knowledge production, circulation, and sharing, and what potential outcomes they would produce.

Given this approach, the key stipulation for the initial concept note was that any proposal must connect with one or a combination of the four themes for which we wish to gather further empirical observations. The themes pertained to:

1. Understanding the various actors' motivations (including incentives and ideologies) for participating in open collaboration;

2. Identifying the enabling infrastructures and technologies for Open Science;
3. Identifying the communities of practice in Open Science in the Global South context; and
4. Documenting the various outcomes of open research practices. This was not restricted to documentation of positive outcomes of Open Science, but also focused on identification of the risks, the negative dimensions, and unanticipated consequences of open research practices.

We asked the partnering projects to formulate research questions around these themes and began to collect much-needed empirical observations and conceptual framings to fill in some major gaps. These include gaps in observations of Open Science practices from the Global South contexts, gaps in conceptualization of "openness" beyond the market-driven and utilitarian framing of open research, gaps in our understanding of knowledge production in a truly equitable and participatory manner, and gaps in policy making pertaining to the support and recognition of Open Science. Specifically, what is the nature of "openness" and its linkage to innovation for the public good, and how can this understanding help formulate and support enabling policies?

Organization of the Book

Despite the relatively short funding period (three years for the coordinating team and two for each sub-project), a number of overlapping themes emerged. Thus, the involvement of "citizens" and non-specialists in the research process and the development of locally specific tools and frameworks for collaboration are common themes for many of the projects (Hillyer et al. 2017). As the coordination team continued to analyze the numerous outputs and final reports from the various projects, further common themes—as well as unique challenges and perspectives—were revealed: these include the power and complexity of multi-actor collaborations, the "situated" nature of openness, openness as a dynamic process of negotiation, and the need for a common language and shared values as the basis for a knowledge commons (Hillyer et al. 2017).

Many of these themes are captured in the OCSDNet Manifesto detailed in Chapter 2. Given the dynamic nature of our understanding of Open Science in general, and of "openness" in particular, the Manifesto is best understood as a "living" document. It will continue to evolve with new inputs, critiques, and interactions with the growing body of literature and the diverse communities from around the world. Detailed technical reports, as well as further OCSDNet–related developments, are posted on the network website.

Given the highly overlapping themes and approaches across projects, the current grouping of the papers into the four sections in this book was somewhat artificial. For example, the issue of governance is present in all the studies, as is the issue of negotiating the boundaries of openness, such that the chapters could have been arranged differently according to the emphasis that we wish to provide. The current groupings were made following the first network-wide meeting in Bangkok in February 2016, at which each team presented a short concept paper based on their research after one year. We asked each of the teams to reflect on what "openness" meant in their local research and community contexts, and we asked them to select the section heading they thought best suited their emphasis. It was generally agreed that the themes of Defining Open Science in Development, the Governance of Open Science Projects, Negotiating Open Science, and the potential roles of Open Science and Social Transformation were sufficiently encompassing. After the meeting, each team continued to develop their original short paper into a full paper that subsequently became the chapters in this volume. As part of the writing process, each chapter was peer-reviewed by other members or authors within the same cluster, and several rounds of revisions ensued. Each paper was also reviewed by the coordination team and by an advisor, who also provided introductory remarks on the section's theme as well as their key takeaways from the papers in each section. The following provides a brief overview of the structure of this collection and details on the four sections, each of which comprises three chapters.

Section 1: Defining Open Science in Development

As noted in Apiwat Ratanawaraha's introduction to this section, the three chapters do not explicitly engage in a formal definition of Open Science or Development. Instead, each project illustrates a form of Open Science in action, involving local actors in addressing a particular issue that was relevant to the community. These studies vividly demonstrate the importance of community members as knowledge makers and how their agency through knowledge production consti-tutes an important form of development. This also echoes Sen's notion of development as freedom, and Appadurai's (2006) call for citizen's right to research—also key components of the Manifesto.

In Chapter 3, "Open Science Hardware (OSH) for Development: Transnational Networks and Local Tinkering in Southeast Asia," Kera and Huang drew from participant observations on a variety of Open Science hardware (OSH) workshops they hosted in Thailand and Nepal. Their work highlights a distinct difference between well-documented understandings of "citizen science" and what they refer to as "little science." They point out that while objectives of conventional citizen science initiatives tend to cater toward larger, institutional, or development objectives, little science affords the opportunity for local participants to engage in tacit reflection, exchange, and tinkering without a firm objective or scientific agenda as the end goal. Under such conditions, the researchers argue that OSH has the opportunity to promote science within everyday activities that are more likely to reflect local realities, as opposed to replicating western constructs or institutionalized forms of science. This work highlights the importance of Open Science beyond the traditional academic environment.

In "On Openness and Motivation: Insights from a Pilot Project in Latin America" (Chapter 4), Lorenzo and colleagues from Colom-bia reflect on their project that aimed to combine the Model Forests (MFs) approach in Costa Rica and Colombia with principles of open "citizen science," environmental conservation, and participatory action research. MFs are social platforms through which diverse groups of stakeholders work voluntarily in partnership toward a common vision of the sustainable development of a given territory or landscape. By bringing community members and academic researchers into spaces of collaboration, the project investigated, among other things, varying levels of motivation toward Open Science for both parties. As a re-sult of various workshops, seven locally driven community initiatives were devised around the theme of local climate change adaptation, including a farming agroecology network, rainwater harvesting pro-gram, tree nursery, and an ecotourism awareness initiative. The level of engagement and high enthusiasm shown by the participants were among the most welcomed aspects of this project.

In Chapter 5, "Contextualizing Openness: A Case Study in Water Quality Testing in Lebanon," Talhouck, Saliba, and a team of environ-mental scientists from the American University in Beirut describe how they engaged citizen scientist volunteers (predominantly women) to explore whether open and collaborative science could be used as an opportunity for environmental management and local development. Using data from a participatory mapping activity, fifty villages were selected that had identified "water quality" as a key area of concern. Local citizen scientists were then trained by the research team to conduct water-quality testing. After rounds of collecting water samples and analysis, researchers found that volunteers were more informed about local water issues, more likely to voice their concerns to political representatives and, hence, to take increased ownership over their community's health and well-being.

Section 2: Governing Open Science

In the introductory remarks for this section, Cameron Neylon reminds us that governance issues related to collaborative community projects are often left unaddressed until problems arise. It is therefore important at the outset to be intentional about trust building, formalizing agreements, ensuring a common language and shared values, and, above all, establishing a clear understanding of who has control over what. The chapters in this section illustrate how these complex dynamics and often conflicting demands play out across different institutional, social, and policy domains.

Chapter 6, "Brazil's Virtual Herbarium, an Infrastructure for Open Science," by Canhos et al. describes an e-infrastructure project known as the Virtual Herbarium. This large distributed network allows for small and large biological collections from across Brazil to compile and share data for increased academic and public access to rich Brazilian botanical records. This project sought to determine who is using this data and for what purposes, as well as to understand the institutional benefits of data sharing. The project reveals many of the benefits and complexities of scientific collaboration and governance issues across institutions and between disciplines while revealing the importance of building Open Science infrastructures in participatory ways. An important lesson learned in this project is that it was important for key participants to have some degree of power regarding their contributions to maintaining the herbarium, particularly with regard to the degree of openness of their data while also having appropriately defined roles that allowed for efficient, longer-term planning and governance of the infrastructure. Communication, transparency, and participation, according to the team, were indispensable for building trust, understanding, and ownership among all actors.

The challenge of working across institutions is also a key theme of Chapter 7, "Collaborative Development of an Open Knowledge Broker for Disaster Recovery Planning," by McNaughton and Rao-Graham. Given the common Caribbean vulnerability to and experience with natural disasters, there is a shared interest and strong regional commitment to collaboration around comprehensive disaster management and the sharing of knowledge resources, artifacts, and response coordination. However, Disaster Recovery Plans (DRPs) are costly but necessary for Small Island Developing States (SIDS) that are frequently affected by hurricanes and earthquakes. Using a "Design Science" approach, this project has sought to develop an Open Source Artifact that could streamline disjointed vocabulary and processes for disaster management between countries and across diverse stakeholders in the region. While revealing the complexities of creating open and enabling infrastructures, this project highlights that the social dimensions of building such tools are key to their long-term success. In that way, the successes of infrastructure should not be based on just their "open" design, but on the longer-term outcomes and social relations between partnering institutions that they facilitate.

When public universities partner with commercial industries for research purposes, there is the potential for great synergies but also for ideological conflict. Chapter 8 by Bolo et al. on "Harmonization of Open Science and Commercialization in Research Partnerships in Kenya" highlights the simultaneous growth in pro-Open Science policies and an increased pursuit of knowledge patents among Kenyan universities and research institutions. Thus, this project sought to assess the national and institutional policy context for the potential of Open Science, and what this shift could entail for partnerships between public and private entities and in trust building. Through an assessment of three case studies, the project concludes that while the country has strong policy guidance around the importance of Open Science and access, the nitty-gritty details of "who owns what" remain an obstacle for true collaboration between institutions and across sectors.

Section 3: Negotiating Open Science

Hebe Vessuri provides the introductory framing for this section. She reminds us that openness is not an end itself, and that in thinking of openness we have to think about the various stages of knowledge production and circulation.

Openness at the knowledge creation stage, the access stage, and the use stage are very different, requiring different actors, capacities, and institutional commitments. The more researchers engaged in the opening process, the more capabilities and tools they will need in support of their work. However, these are not currently being provided by scientific institutions or policy schemes, particularly in the Global South, where there are virtually no models that inform how to build good practices of openness at the laboratory level. These gaps will need to be addressed by policy makers who wish to see greater adoption of open practices.

Collaboration in scientific knowledge production has been historically dominated and driven by hegemonic (Northern) countries, while non-hegemonic countries tend to take on secondary roles. Nonetheless, the growing discourse on Open Science provides the opportunity to reflect critically on the roles and outcomes of collaborative knowledge creation in Global South contexts. In Chapter 9, "Co-production of Knowledge, Degrees of Openness, and Utility of Science in Non-hegemonic Countries," Ferpozzi and a diverse research team draw on four in-depth case studies throughout Latin America, focusing on neglected socio-scientific topics that are of importance to local communities, but may not be viewed as worthy of investigation by mainstream knowledge makers (e.g., pharmaceutical companies) due to their low-profit potential. Through their analysis, the team identified that drivers—that is the individuals or groups initially engaged in mobilizing scientific knowledge for particular outcomes— are the keys to gauging the anticipated degree of openness within processes of knowledge production. These four case studies illustrate that the degree of openness of knowledge produced from research is dependent on the kinds of research being performed, who drives the research agenda, and, importantly, for whom the research is being performed. Thus, openness is situated and highly conditional on the conditions of knowledge production.

Chapter 10 by Traynor et al. on "Tensions Related to Openness in Researching Indigenous Peoples' Knowledge Systems and Intellectual Property Rights" further explores issues of boundaries in practices of Open Science, focusing particularly on research with Indigenous peoples in South Africa. The authors examine the colonial notions of "science" and "openness" and how historical injustices and lack of redress influence the context in which current research is situated. This project broadly aimed to develop a political, ecological approach to understanding the relationship between climate change, intellectual property, and Indigenous peoples. The approach taken was influenced by "decolonizing methodologies" and feminist perspectives and, like other projects in the network, employs participatory action research methodologies to guide not just the substantive but also procedural elements of the research. The authors share their experience with developing "community-researcher contracts" in an attempt to make researchers more accountable to Indigenous Nama and Griqua communities and to adequately protect their Indigenous knowledge. They recount the challenges of negotiating the contracts and how they conceptualized the concept of a "situated openness"—a way of doing research that assumes knowledge production and dissemination is situated within particular historical, political, socio-cultural, and legal relations.

In "Negotiating Openness in Science Projects: Case Studies from Argentina" (Chapter 11), Arza and Fressoli present their project, which analyzes four diverse cases of Open Science in Argentina, characterizing what is being opened, how, and who participates in these practices. Their study suggests that as scientists progressively open more stages of their research, they enter into a social terrain that challenges their formal scientific norms and ways of working. This process of transition also puts new strains on Open Science practitioners, as each stage may entail a new form of contradiction and, hence, negotiation with traditional institutional norms and structures. These moments are studied through the notion of "boundary objects" to understand how scientists negotiate meanings, tools, and several forms of communication with actors from outside the laboratory. The chapter concludes by suggesting that there is a need to identify and build exemplary cases of Open Science that allow for the construction of good practices.

Section 4: Open Science for Social Transformation

Halla Thorsteinsdóttir's introduction provides an overview of how three very different grassroot projects offer insights into how Open Science practices and, in particular, knowledge co-production could have transformational effects, potentially leading to a process of shifting institutionalized power relationships, norms, values, and hierarchies over time.

In Post-Soviet Kyrgyzstan, "science" is understood by most citizens to consist of highly technical and expensive activities to be performed by scientific "experts." The Kyrgyz Mountains Environmental Education and Citizen Science (KMEECS) project sought to challenge these widely held assumptions by engaging rural school children and their teachers in biological, chemical, and physical analyses of water quality, as well as water flow measurement and mapping of locally relevant water resources. Rosset et al. recount their study design and key results in Chapter 12, "Experimenting with Openness as a Seed for Social Transformation: Linking Environmental Education and Citizen Science in Remote Mountain Villages of Kyrgyzstan." Using a participatory action research approach, this project looks at the transformational potential of citizen science initiatives for environmental monitoring and education. It also provides insight on the motivational factors related to citizen science at the local level and the complexities of collaboration and support between community and governmental institutions in a post-Soviet state.

In Chapter 13, "Open Science and Social Change: A Case Study in Brazil" Albagli and a diverse research team raise fundamental questions about openness and its practice. The community of Ubatuba in São Paulo, Brazil, is located in a dense rainforest region with a diverse mix of Indigenous communities, researchers, activists, and policymakers interested in the area. It makes a compelling case study for examining the potential of Open Science from a sustainable development perspective. This project draws on a reflective, action-oriented research approach to understand the institutional, cultural, and political challenges involved in the adoption of an open approach for development in Ubatuba, Brazil, by interacting with a variety of different communities and actors. The authors conclude that, on the one hand, open and collaborative science does create new spaces and methods for traditionally marginalized groups to engage in scientific discussions and local problem-solving, mainly in controversial and conflict situations and as a condition for resilience and political struggle for alternative paths of development. On the other hand, the very idea of openness is under dispute: what (open) science and for whom? The idea of science itself is also under dispute, and nowadays this dispute lies at the very core of democracy building.

Further questioning the notion of for whom and by whom is science being opened, a diverse Francophone team led by Piron has been working on "Towards African and Haitian Universities in Service to Sustainable Local Development: The Contribution of Fair Open Science" (Chapter 14). Having identified the historically unjust and devastating legacy of colonialism and its impact on higher education throughout Francophone Africa, the team sought to define and promote Open Science and Open Access in French-speaking West Africa and Haiti using a network-building and advocacy approach, using social media tools, surveys, and workshops. Targeting the lack of access to academic journals experienced by many institutions within these regions, the team engaged university students and staff in discussions about access to research and the proportional lack of representation of Southern (and particularly French-speaking African and Haitian) researchers in the production of scientific knowledge. This group has also been forcefully promoting the concept of "cognitive justice" within and beyond the network—a concept that acknowledges the right of human beings to participate in the creation of knowledge that is relevant to their own lives, experiences, and ways of knowing.

The idea of cognitive justice resonates highly with other projects in the network, and it constitutes one of the seven principles of the OCSDNet Manifesto set forth in the Introduction to this volume (see Chapter 2), where we provide details of the consensus-building process, the background to each principle, as well as the key sources for the observations and inspirations behind each principle.

Concluding Remarks

One of the key network findings is that there is no single or universal concept of Open Science that is sufficient to encompass the diversity of knowledge traditions and practices from around the world. Hence the term Open Science and the notion of "openness" is highly situated, constantly subjected to negotiation according to local contexts and historical contingencies. Our collective observations therefore challenge the tendency to define Open Science as a set of technical infrastructure, workflow, protocols, and licensing conditions that can be universally applied regardless of context, history, and human agency.[6]

Such a tendency mirrors the Eurocentric tradition of seeing Western Science as universal and superior, while rendering invisible other forms of knowing that are deemed unscientific because they do not fit into a monolithic view of how science is defined (Shiva 2016). This tendency also reflects the reality that global processes of knowledge production and research agenda setting have historically been shaped and solidified by a set of privileged, powerful, and exclusive actors and institutions, ultimately influencing the way in which the world understands "valid" and "legitimate" scientific knowledge and research agenda (De Sousa Santos 2014). This limited representation of knowledge leads to an incomplete understanding of the world and of the issues affecting local communities (Sillitoe 2007; Moletsane 2015). It also leads to what David Hess (2016) refers to as "Undone Science," namely "areas of research that are left unfunded, incomplete, or generally ignored, but that social movements or civil society organizations often identify as worthy of more research" (Frickel et al. 2010, 444). Unchallenged, this neocolonial, market-driven system will continue to exacerbate knowledge and research inequalities with serious consequences for sustainable and equitable development (Hall et al. 2014; Hall and Tandon 2017b; Fuchs 2017).

One of the goals of this book was to identify the structural, technical, policy, and cultural contexts for Open Science among the twelve projects in order to begin to recognize the plurality and diversity in the framing and meanings of "science," "openness," and "development." We believe the case studies provide a range of critical discussion and reflection on the nature of openness and its implications for knowledge production while looking ahead to suggest how these ideas could be better studied and applied to make Open Science principles and practices more inclusive and relevant to local development challenges.

Throughout this book, readers will encounter different examples of how "openness" cannot be simply taken for granted or assumed to be universally good, as the notion can just as easily be used as a tool to dispossess others' knowledge and to enrich those who are already powerful and well-resourced. Openness as a concept must therefore be rooted in proper and historical and political contexts, otherwise we risk replicating the power inequality and asymmetry that we seek to challenge and replace (Christen 2012; Moletsane 2015; Gurstein 2015; Cronin 2016). It is therefore important to ask for whom "science" is being opened, by whom, who stands to benefit, and who may suffer the risks of being further excluded and marginalized. Such a call is one of the most consistent themes throughout this volume.

We hope this book will also stimulate further research and debates on how best to collectively design knowledge systems, including production and dissemination infrastructure that are not only open, but are inclusive and equitable for all, while fostering dialogues with multiple voices and nourishing diverse ways of knowing, knowledge representations, and, more importantly, their legitimization. Openness may be necessary, but it is not sufficient for substantive structural or transformative changes to occur.

Toward this goal, this book is an invitation for readers to imagine what Open Science may look like when viewed through the lens of diverse cultures, epistemologies, research traditions, disciplinary background, and, more importantly, through critical decolonizing lenses that question the history and power structures of global knowledge-making institutions, particularly those vested with the authority and power to produce, legitimize, and circulate knowledge to maintain their status quo (Connell 2007; Mignolo 2011; Czerniewicz 2015; Hall and Tandon 2017a).

The richness and diversity of perspectives, institutional settings, and local actors represented by the twelve chapters in this book are truly impressive. Our hope is that the many new observations stemming from these studies will begin to fill in the conceptual and empirical gaps in the literature, and, more importantly, policy gaps that directly affect resource allocation and future research. But these gaps remain large, and much work remains to be done. In the process of presenting these studies, we trust we will stimulate further debates and critical dialogues on what openness means for knowledge making and circulation in various contexts. Most importantly, we hope the questions of "whose open science?" and "for whom is science being opened?" will continue to be raised. These are critical questions as many of the lofty goals of sustainable development cannot be easily achieved without acknowledging the importance of epistemic or cognitive justice as the foundation for development. In the process, we are also modelling Open Science on what Connell (2018: 404) referred to as "mosaic epistemology," which "offers a clear alternative to northern hegemony and global inequality, replacing the priority of one knowledge system with respectful relations among many."

Notes

1. We use this term to denote regions that are historically and structurally excluded from institutionalized networks of power, authority, visibility, and access in global knowledge production. These regions span across Africa, Asia, the Americas, as well as Europe.
2. See biographical sketches of the advisors at https://ocsdnet.org/about-ocsdnet /the-team/.
3. See the original Call for Concept Notes at https://ocsdnet.org/application-2/.
4. See the distribution map of the projects at https://ocsdnet.org/ocsdnet-projects/.
5. For an interactive view of the key research areas and geographic locations of the twelve projects, see https://ocsdnet.org/ocsdnet-projects/.
6. For example, the highly cited definition in the OECD document (2015) and the definition by FOSTER (https://www.fosteropenscience.eu/content/what-open -science-introduction), and EU-funded project on training for Open Science.

References

Appadurai, Arjun. 2006. "The Right to Research." *Globalisation, Societies and Education* 4 (2): 167–77.

Bartling, Sönke, and Sascha Friesike, eds. 2014. *Opening Science: The Evolving Guide on How the Internet is Changing Research, Collaboration and Scholarly Publishing*. Cham: Springer International Publishing. http://link.springer .com/10.1007/978-3-319-00026-8.

Bollier, David, and Silke Helfrich, eds. 2014. *The Wealth of the Commons: A World Beyond Market and State*. Amherst: Levellers.

Chan, Lelsie, and Eve Gray. 2014. "Centering the Knowledge Peripheries through Open Access: Implications for Future Research and Discourse on Knowledge for Development." In *Open Development: Networked Innovations in International Development*, edited by Matthew L. Smith and Katherine M.A. Reilly, 197–222. Cambridge, MA and Ottawa: MIT Press and IDRC. https://idl-bnc-idrc.dspacedirect.org/bitstream/handle /10625/52348/IDL-52348.pdf.

Chan, Leslie, Angela Okune, and Nanjira Sambuli. 2015. "What is Open and Collaborative Science and What Roles Could It Play in Development?" In *Open Science, Open Issues*, edited by Sarita Albagli, Maria Lucia Maciel, and Alexandre Hannud Abdo, 87–112. Brasília: Instituto Brasileiro de Informação em Ciência e Tecnologia (IBICT). http://hdl.handle.net/1807/69838.

Charmaz, Kathy, and Linda Liska Belgrave. 2015. "Grounded Theory." In *The Blackwell Encyclopedia of Sociology*, edited by George Ritzer. Hoboken. NJ: Wiley-Blackwell. https://doi.org/10.1002/9781405165518.wbeosg070.pub2.

Christen, Kimberly A. 2012. "Does Information Really Want to be Free? Indigenous Knowledge Systems and the Question of Openness." In *International Journal of Communication*, 6 (0): 24. http://ijoc.org/index.php /ijoc/article/view/1618.

Connell, Raewyn. 2007. *Southern Theory: The Global Dynamics of Knowledge in Social Sciences*. Malden, MA: Polity.

———. 2018. "Decolonizing Sociology." *Contemporary Sociology*, 47 (4): 399–407. https://doi.org/10.1177/0094306118779811.

Cronin, Catherine. 2016. "Openness and Praxis (at #SRHE)." https://catherine cronin.wordpress.com/2016/11/28/openness-and-praxis/.

Czerniewicz, Laura. 2015. "Confronting Inequitable Power Dynamics of Global Knowledge Production and Exchange: Feature – Opinion." *Water Wheel* 14 (5): 26–28. http://journals.co.za/content/waterb/14/5/EJC176212.

Frickel, Scott, Sarah Gibbon, Jeff Howard, Joanna Kempner, Gwen Ottinger, and David J. Hess. 2010. "Undone Science: Charting Social Movement and Civil Society Challenges to Research Agenda Setting." *Science, Technology, & Human Values*, 35 (4): 444–73. https://doi.org/10.1177/0162243909345836.

Friesike, Sascha, Bastian Widenmayer, Oliver Gassmann, and Thomas Schildhauer. 2015. "Opening Science: Towards an Agenda of Open Science in Academia and Industry." *The Journal of Technology Transfer*, 40 (4): 581–601. https://doi.org/10.1007/s10961-014-9375-6.

Frischmann, Brett M., Michael J. Madison, and Katherine J. Strandburg, eds. 2014. *Governing Knowledge Commons*. Oxford: Oxford University Press.

Fuchs, Christian. 2017. "Critical Social Theory and Sustainable Development: The Role of Class, Capitalism and Domination in a Dialectical Analysis of Un/Sustainability." *Sustainable Development*, 25 (5): 443–58. https://doi.org/10.1002/sd.1673.

Glaser, Barney and Anselm Strauss. 1967. *The Discovery of Grounded Theory: Strategies for Qualitative Research*. Chicago, IL: Aldine.

Gurstein, Michael. 2015. "Why I'm Giving Up on the Digital Divide." *The Journal of Community Informatics* 11 (1). http://ci-journal.net/index.php/ciej /article /view/1210.

Harding, Sandra. 2006. *Science and Social Inequality: Feminist and Postcolonial Issues*. Urbana: University of Illinois Press.

Hall, Budd, Cristina Escrigas, Rajesh Tandon, and Jesús Granados Sánchez. 2014. "Transformative Knowledge to Drive Social Change: Visions for the Future." In *Higher Education in the World 5. Knowledge, Engagement and Higher Education: Contributing to Social Change Series*, edited by The Global University Network for Innovation (GUNi), 301–310. Basingstoke, UK: Palgrave Macmillan.

Hall, Budd, and Rajesh Tandon. 2017a. "Decolonization of Knowledge, Epistemicide, Participatory Research and Higher Education." *Research for All* 1 (1): 6–19. https://doi.org/10.18546/RFA.01.1.02.

Hall, Budd, and Rajesh Tandon. 2017b. "Participatory Research: Where Have We Been, Where are We Going?—A Dialogue." *Research for All* 1 (2): 365–74. https://dspace.library.uvic.ca/handle/1828/8562.

Hess, Charlotte, and Elinor Ostrom, eds. 2006. *Understanding Knowledge as a Commons: From Theory to Practice.* Cambridge, MA: MIT Press.

Hess, David J. 2016. *Undone Science: Social Movements, Mobilized Publics, and Industrial Transitions.* Cambridge, MA: MIT Press.

Hillyer, Rebecca, Alejandro Posada, Denisse Albonaz, Leslie Chan, and Angela Okune. 2017. "Framing a Situated and Inclusive Open Science: Emerging Lessons from the Open and Collaborative Science in Development Network." In *Expanding Perspectives on Open Science: Communities, Cultures and Diversity in Concepts and Practices: Proceedings of the 21st International Conference on Electronic Publishing* 18. IOS Press.

Mignolo, Walter. 2011. *The Darker Side of Western Modernity: Global Futures, Decolonial Options.* Durham, NC: Duke University Press.

Moletsane, Relebohile. 2015. "Whose Knowledge Is It? Towards Reordering Knowledge Production and Dissemination in the Global South." *Educational Research for Social Change (ERSC)* 4 (2): 35–47. http://ersc.nmmu .ac.za/articles/Vol_4_No_2_Moletsane_pp_35-48_October_2015.pdf.

Nielsen, Michael. 2011. *Reinventing Discovery: The New Era of Networked Science.* Princeton, NJ: Princeton University Press.

Santos, B. de S. 2014. *Epistemologies of the South: Justice Against Epistemicide.* Herndon: Routledge.

Shiva, Vandana. 2016. *Earth Democracy: Justice, Sustainability and Peace.* London: Zed Books Ltd.

Sillitoe, Paul, ed. 2007. "Local Science v. Global Science: An Overview." In *Local Science vs. Global Science: Approaches to Indigenous Knowledge in International Development*, 1–22. New York: Berghahn Books.

CHAPTER 2

Principles for an Inclusive Open Science: The OCSDNet Manifesto

Denisse Albornoz, Becky Hillyer, Alejandro Posada,
Angela Okune, and Leslie Chan

Abstract

The OCSDNet Manifesto is the result of one year of participatory consultations and debates among members of the Open and Collaborative Science in Development Network (OCSDNet), a network of twelve research-practitioner teams from Latin America, Africa, the Middle East, and Asia. Through research projects grounded in diverse regions and disciplines, OCSDNet members explored the scope and possibilities of Open Science as a transformative tool for development thinking and practice. They offer the Open and Collaborative Science Manifesto as a foundation upon which to reclaim the narrative about what Open Science means and how it can realize a more inclusive science in development. This article outlines the seven principles of the OCS Manifesto, which are grounded in critical development theory and empirical examples arising from OCSDNet research teams. Taken as a collective, this chapter articulates the network's vision of an inclusive and critical understanding of open and collaborative science in the context of development. In doing so, it is our intention to contribute toward challenging homogeneous, decontextualized, and dehistoricized definitions of Open Science, and support calls for a more situated knowledge and an open and collaborative science for well-being, development, and collective prosperity.

Introduction

The development of the Open and Collaborative Science in Development (OCSDNet) Manifesto was largely in response to what we perceived as the lack of transformative and critical approaches to Open Science in the global scientific and development community. Most mainstream narratives about OS, emerging particularly from Europe and North America, envision Open Science as a system of technology-driven tools and processes (e.g., OECD 2015; Grigorov et al. 2015; Schmidt et al. 2016) that, when utilized, are assumed to accelerate scientific discoveries, improve transparency and reproducibility of research, increase research uptake, and improve accountability to the scientific community as well as to the public (Nosek et al. 2015; Leonelli et al. 2015; McKiernan et al. 2016). While we recognize a great deal of progress has been made through technology-enabled collaboration, we also note that the established voices in the Open Science community have failed to address how the current approach to "open" exacerbates and amplifies disparities in knowledge production and circulation (Nyamnjoh 2009, 2013; Tkacz 2012; Tyfield 2013; Kansa 2014; Okune et al. 2016).

For the research teams within OCSDNet, the collective framing of Open and Collaborative Science (OCS) was a highly iterative process of consensus building. Since each research context was highly distinct, and all teams had their own preconceptions of what "openness" should entail (see subsequent chapters), the network consequently spent many hours debating and articulating our respective values around how we should work together in order to practise inclusive Open Science (Albornoz et al. 2017). The outcome was an optimistic, reflective, and critical Manifesto that consolidates the common values, language, and vocabulary used among the OCSDNet community to discuss openness, collaboration, and inclusion in science, resulting in seven principles[1] that are relevant across multiple contexts in the Global South. The intention of the Manifesto is not to offer a prescriptive formula for practising OCS, but rather it seeks to acknowledge the collective values that we share, as influenced by experience conducting empirical and action research within the network.

In particular, network members collectively questioned and discussed the configuration and roles of structural power in their contexts, asking: To whom does knowledge belong? Who benefits from the production and circulation of scientific knowledge? Who gets to participate in knowledge production processes? And, in what ways can technology be used to increase the agency of more people over scientific knowledge production?

Using these questions as the starting point for deliberation, network members came to agree on a set of seven principles (see Figure 2.1) that are relevant across multiple contexts in the Global South. We propose that Open and Collaborative Science in Development:

1. Enables a knowledge commons where all individuals have the means to decide how their knowledge is governed and managed to address their needs;
2. Recognizes cognitive justice and the need for diverse understandings of knowledge making to co-exist in scientific production;
3. Practises situated openness by addressing the ways in which context, power, and inequality condition scientific research;
4. Advocates for each individual's right to research and enables different forms of participation at all stages of the research process;
5. Fosters equitable collaboration between scientists and social actors, and cultivates co-creation and social innovation in society;
6. Incentivizes inclusive infrastructures that empower people of all abilities to make and use accessible open-source technologies; and
7. Uses knowledge as a pathway to sustainable development, equipping every individual to improve the well-being of our society and planet.

Methodology of Co-constructing a Manifesto

The idea of constructing a Manifesto was born in May 2015, after several members of the network met in Singapore to present at the ICTD Conference 2015. There we realized the network needed to produce a document that outlined our position in the Open Science debate, reflecting our commitment for a more inclusive, collaborative, and just approach to knowledge production. While network members came from different disciplinary, cultural, and ethnolinguistic backgrounds, we shared the concern that the mainstream narrative of Open Science needed to be reclaimed and reimagined, from the technocentric rhetoric dominating the debate to a set of common values that promote the social embeddedness of knowledge at all levels of society.

Figure 2.1. OCSDNet Manifesto Infographic

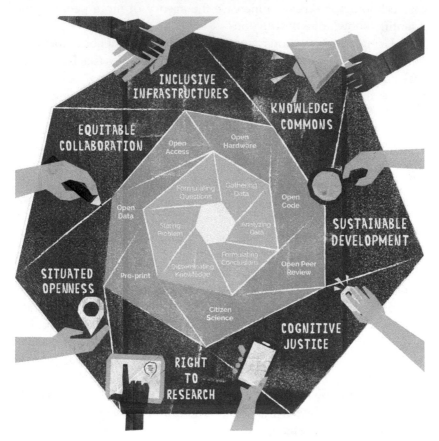

From June 2015 onward, the OCSDNet coordination team conducted a series of participatory, collaborative, and horizontal consul-
tative processes, which took place over the course of one year, to tap
into the synergies and divergences in our vision for Open Science.
These included detailed analysis of formal project reports and position papers submitted by each team, as well as more informal group
calls, workshops, and collaborative editing sessions in which network
participants shared and debated their views about what Open Science means for them and their communities. During the process of
consultation, the coordinating research team specifically looked for
common keywords, themes, and ideas that encapsulated the principles
and processes guiding the research practice of the twelve research

teams. In addition, we also carried out feedback sessions to improve the content of our document and to develop a tone, language, and dissemination format that reflects the inclusive and collaborative spirit of the scientific model it proposed. The result was a reflective and critical Manifesto that we hope will promote conversation in the scientific community and beyond about the need for an expanded and more inclusive definition of Open Science.

It is important to acknowledge that the process of consultation and the framing of this Manifesto were informed by the many scholarly traditions that have historically challenged the hegemony of positivism and a market-driven scholarly communication system. As such, many of the ideas behind the principles comprising the Manifesto are not new and have been central to fields such as critical theory, postcolonial, feminist, and Indigenous epistemologies among others (Figure 2.2). As part of our process, we gathered these various ideas and documented the ways in which they informed the principles of the Manifesto in a collaborative, annotated bibliography and reading

Figure 2.2. OCSDnet Manifesto Principles and Reading List of Key Authors

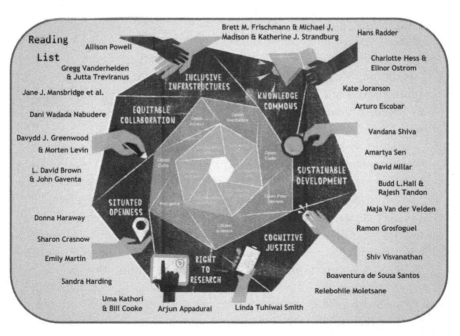

list.[2] Through this open resource, we aim to pay homage to the work of the many authors, but also to further visualize the intersections between Open Science and the many streams of social justice scholarship. We also hope that its users will continue to make suggestions and contribute to it as the understanding of Open Science and the field continues to expand.

The following section provides expanded details on each of the principles comprising the Manifesto.

Toward an Inclusive Open Science Through the OCS Manifesto

Principle 1: Knowledge Commons

> A knowledge commons is established when intellectual and cultural resources are collectively managed, shared, used, and governed by all or most members of a community.

The conceptual framework that initially inspired the creation of research questions to guide the work of OCSDNet was based on the Institutional Analysis and Development (IAD) framework developed by Elinor Ostrom and colleagues over several decades of work on natural resource commons and their governance. Ostrom's work challenged the conventional wisdom about the need for government regulation of public resources in order to attain sustainability and benefit sharing (Ostrom 1990, 2005). The IAD framework has been applied to a variety of studies on how people collaborate and organize themselves across organizational and state boundaries to manage common resources such as forests and fisheries, which often cross or flow through national boundaries (Ostrom 1990, 2005).

In the context of the commons, OCS offers potential opportunities for increasing diverse forms of participation in the circulation and construction of scientific knowledge that have traditionally excluded actors from outside powerful and wealthy research institutions. The diversity of participation and the integration of community actors allow for scientific research that lends itself more easily to addressing local, context-specific, development issues. It is this potential to form collaborative connections across traditional and institutional boundaries, we argue, that is the key feature and attraction of OCS, particularly for those who have been historically excluded. In the long term, OCS may lead to structural transformation of knowledge institutions and cultural changes that democratize the benefits of science for all, not just for the elites. This further raises important questions around collective governance, inclusive participation, and sustainability in relation to maintaining a knowledge commons (Frischmann et al. 2014; Bollier and Helfrich 2014).

Taking into account the unique attributes of knowledge and information that are distinct from natural resources, Frischmann et al. (2014) have more recently modified the IAD framework into a Knowledge Commons framework to aid other researchers with empirical research on different forms of commons. The framework provides a number of guiding research questions about the nature of the community in question, the kind of resources in use, existing institutional arrangements, and interactions that take place within the community. In recent years, a number of researchers have applied the modified IAD framework to study a variety of "knowledge commons," from Open Source Software and the SourceForge repository (Schweik and English 2012) to genomic commons (Van Overwalle 2014) to the well-known Galaxy Zoo citizen science crowdsourcing project (Madison 2014).

Given the diversity of commons, it is not surprising that there is not a fixed set of rules for developing a knowledge commons. Instead, Hess and Ostrom (2005: 53) reminded us that

> the rules connected with knowledge, epistemic communities, and information technologies must continually be adapted as those technologies and communities change and grow. Rules need to be flexible and adaptable in order to create effective institutional design and ensure resource sustainability.

A case study from the network could serve to illustrate this point. The OCSDNet project in Brazil, "Virtual Herbarium as OCS Infrastructure," has been involved in the design and governance of a "Virtual Herbarium," consisting of a consortium of large and small Brazilian institutions, all of which agreed to centralize their botanical records within an openly accessible e-database for improved access by researchers and the general public (see Chapter 6). The intention of the project was to understand who was accessing the botanical records and for what purposes. As a whole, the project uncovered surprisingly high rates of access to records within the centralized system, particularly in comparison to access at the individual-institutional level. However, despite this increased overall usage of the botanical data, the team was surprised when one of the larger institutions withdrew their participation and all of their respective data from the consortium. The institution's assumption was that as a large and well-resourced research institution, they had previously been a key gatekeeper of botanical records. But now that access had been made more readily available for smaller institutions, there was an understanding that the larger institution's "status" had been somewhat diminished. This example illustrates the tension that can be created when institutions of varying power participate in a common project. In this case, the larger institution did not feel they were receiving sufficient return for their participation, while the smaller institutions were benefiting more from participation in the Virtual Herbarium.

The fear of "free rider" or unequal benefits is often a disincentive for individuals or organizations to participate in common pool resources and collective action (Ostrom 2009). Nonetheless, this example also highlights the opportunity for often-marginalized actors to benefit from the development of a knowledge commons, through increased agency to access, participate, and govern the creation of knowledge. This case demonstrated how the creation of a knowledge commons is not a straightforward process, but indeed involves iterative debate and reflection on how existing power structures, hierarchies, and the cultures of collaboration make and shape the way that institutions operate. As Hess and Ostrom (2005) acknowledge, these negotiations become even more important as new technologies open up increasing opportunities for diverse forms of collaboration and resource sharing.

Principle 2: Cognitive Justice

> This ideal considers that all individuals and communities, regardless of their culture, gender, socioeconomic status, or language, should be able to fully exercise their capabilities to use, share, and create knowledge. It recognizes the diversity of ways of knowing and plurality of knowledge and fosters the interaction of diverse knowledge traditions.

Principle 2 of the OCS Manifesto acknowledges that both presently and historically Western-centric knowledge, traditions, research practices, and institutional power structures have largely defined what is and what is not considered to be a legitimate way of understanding the world (Mignolo 2002; Grosfoguel 2007). Many mainstream scientists are taught to pursue a positivist scientific methodology with the intention of arriving at a singular and objective truth (Mies and Shiva 2014). While positivism does serve specific purposes, feminist scholars of science (e.g., Harding 1986, 2006, 2015; Haraway 1988; Shiva 1995, 2016) have been exposing methodological and gender biases in western science for decades.

Against the idea that science is neutral and objective, Harding developed the "standpoint theory," suggesting that one's perspectives are grounded and shaped by his or her social and political experiences (Harding 2015). Thus, the way that one understands the world and, hence, subscribes to a particular version of legitimate knowledge is to a large degree dependent on lived experiences, or a personal standpoint. This theory is further reinforced by what Harding calls "strong objectivity"—the notion that the lived experiences of individuals (particularly those who tend to be politically or economically marginalized) are useful for developing more objective accounts of the world in which they live. In other words, as grounded individuals, their reflections are often more acute and accurate, as opposed to the skewed and episodic observations of outside researchers who often parachute into an artificial research setting that may not be grounded in local reality.

Harding's philosophy echoes the growth of Indigenous networks and decolonizing movements around the world who are calling for cognitive justice and epistemic diversity in science and development. The movements of the Andean highlands people succeeded in their push to incorporate the Indigenous philosophy of *Buen Vivir* into the constitutions of Bolivia and Ecuador in 2006–7 (Gudynas 2011; Monni and Pallotino 2015). Latin American scholars of decolonizing studies—notably Arturo Escobar (2011), Boaventura de Sousa Santos (2008), and Walter Mignolo (2011)—have been calling for another world of decolonized science as an alternative to northern epistemologies. They likewise assert that social justice is not possible without cognitive justice. Santos (2014) has gone further to suggest that Western science and scientific enterprises commit "epistemicide" when the knowledge and experiences of the majority of the world's peoples are disregarded and devalued.

It is of significance that these calls for diverse epistemologies and cognitive inclusion are echoed by many of the projects in OCSD-Net. For instance, OCSDNet's *Projet SOHA* "OCS, Empowerment and Cognitive Justice"—a networked collaboration stretching across French-speaking West Africa and Haiti—focuses on raising awareness around the cognitive injustices that many university students in the region are likely to encounter over the course of their studies (see Chapter 14). Along with some of the more obvious technical limitations to accessing knowledge (such as a lack of Internet connectivity, computers, electricity, etc.), the project has also uncovered evidence to suggest that many institutions in West Africa tend to subscribe to and promote many of the same norms and standards around knowledge creation and legitimacy as one might find in Northern institutions. In doing so, institutions may intentionally (or unintentionally) delegitimize forms of knowledge that do not conform to these norms, such as the use of oral traditions, arguments drawn from Indigenous worldviews, or alternative forms of publishing.

From a development perspective, these Northern-centric learning cultures are potentially harmful, as they tend to promote and idealize de-localized and imposed forms of knowledge and research, rather than prioritizing local solutions to local challenges. Using a network-building and information-sharing approach, the intention of *Projet SOHA* has thus been to foster a culture of "science aimed at the creation of locally relevant, freely accessible, and reusable knowledge by empowered and confident researchers using not only epistemologies from the North, but all kinds of epistemologies and methods." From their work, they have found that young Haitian and West African scholars have a strong willingness and key role to play in establishing a culture of science and learning that is inclusive of a diversity of worldviews and which is intent on solving complex, local development issues.

Principle 3: Situated Openness

> A concept that assumes knowledge is situated within particular historical, political, and socio-cultural relations. It addresses inequalities and hierarchies of knowledge production and its inherent conflicts.

Largely borrowing from the work of feminist scholars, Principle 3 of the OCSDNet Manifesto recognizes that knowledge is situated within a very particular socio-cultural context and, hence, the importance or legitimacy of that knowledge may be limited to those individuals who are impacted by similar circumstances. By looking at knowledge in this way, there is a tendency to centre knowledge as inherently personal in nature, in opposition to neoliberal philosophies that tend to promote knowledge hierarchies that separate data from human needs and local challenges. As Haraway (1988) notes: "Feminism loves another science: the sciences and politics of interpretation, translation, stuttering, and the partly understood [...] Feminism is about a critical vision consequent upon a critical positioning in un-homogeneous gendered social space. Translation is always interpretative, critical, and partial" (589).

OCSDNet teams have grounded their research on understanding the way that scientific knowledge is situated within particular localized circumstances with the intention of solving complex local challenges. Within the network, the concept of "situated openness" was brought to the forefront through the work of the Natural Justice research team in South Africa (see Chapter 10). In this case, the research team found that indigenous communities with whom they had worked expressed a clear lack of trust around the idea of sharing their knowledge with scientists or outside researchers due to past instances where generational knowledge had been appropriated and/ or commodified without consent, credit, or compensation to their communities. This sentiment was echoed by the Argentinian research team involved in the "Negotiating Openness in Science Projects" (see Chapter 11), which focused on exploring alternative spaces of knowledge production for social movements. In one of the Manifesto consultations, the team raised the following point:

> The idea that openness is always for the better, should be revised and contextualized. That idea is [especially] a hard sell when your [community] is being harassed by the government, the academic establishment, or political actors. While openness can be a means for empowering and strengthening alternative science, if wrongly used, it might become an effective means to weaken or destroy it.

In both instances, the very notion of "openness" is being questioned. An inclusive OCS should not imply "openness for all," but rather our findings suggest that a situated perspective must be taken to ensure that openness is fair to those involved and grounded within a context that is cognizant of the historical experiences and present-day constraints of marginalized actors. These findings thus challenge the common Open Science rhetoric that tends to imply that openness is always good or desirable for all.[3]

Practising both situated openness and cognitive justice within a feminist framework helps us to understand the power relations, inequalities, and structural constraints that shape the way knowledge is produced, legitimized, and adapted, as well as to imagine the types of frameworks and tools that could be used to enable all social groups to define the conditions under which their knowledge can be shared and used. For instance, in the case of the Natural Justice project, the team changed the course of their research activities toward the creation of a community-generated research contract that would allow community members themselves to define the ways in which their knowledge should be used and protected during negotiations with external researchers.

Importantly, the network has recognized that the inclusion of diverse actors and diverse epistemologies is not merely a goal to be attained, but a process of constant negotiation and reflection, of understanding power relations and group dynamics, and intentionally reconfiguring research methodologies and practices to address the knowledge needs of those who are often marginalized from the research process. This is facilitated through the use of digital tools and processes, but can only be achieved through engagement in respectful debate, discussions, and the co-creation of meaningful, collective knowledge (Hall and Tandon 2017; Brown and Gaventa 2008).

Principle 4: Right to Research

> The right of individuals to participate at all stages of the research process as a means to gain strategic knowledge about their communities and fulfill their capabilities.

An individual's "right to research" is a concept first theorized by Arjun Appadurai (2006), and that has largely come to shape the way that OCSDNet members understand and define the connection between OCS and development. Appadurai (2006) suggests that development should be defined as "the capacity to aspire" (176) or, in other words, an individual's ability to dream, set goals, and achieve them. He further suggests that poverty and inequality are "the uneven distribution of this capacity" (Appadurai 2006: 176). Appadurai's ideas are similar in nature to Amartya Sen's capabilities approach (2003), suggesting "full citizenship…requires the capacity to make strategic inquiries… and gain strategic knowledge on a continuous basis" (168). Thus, the ability to access and create locally grounded, contextually relevant knowledge is, in Appadurai's view, a foundational component for human development. He calls for all human beings to claim "the right to the tools through which any citizen can systematically increase that stock of knowledge which they consider most vital to their survival... and...claims as citizens" (Appadurai 2006: 168).

This understanding has become central to OCSDNet's conceptualization of an inclusive OCS. In particular, it allows for the recognition that access to knowledge is necessary, though not sufficient, as the processes of creating and sharing knowledge are likewise important for the formation of an inclusive OCS. Appadurai's work henceforward not only pushes us to consider who is involved in collecting data for Open Science, but also raises questions around who is designing the research questions, methods, and processes of data analysis.

The concept of "citizen scientist" becomes important here, and it has been popular within many mainstream discourses around Open Science. In many circles, a citizen scientist is often interchangeable with a data-collection volunteer. For instance, Silvertown (2009) refers to a citizen scientist as "a volunteer who collects and/or processes data as part of a scientific enquiry," while Cohn (2008) defines them as "volunteers who participate as field assistants in scientific studies." Although these forms of citizen science may have important outcomes for knowledge production and development, there tends to be less focus on the individual as a local expert, or co-researcher, who is able to have input in the design of the research process, questions, and data analysis.

For this reason, OCSDNet is cognizant that personal agency must be deeply entrenched within an inclusive OCS, and the consequent distribution of power within processes of knowledge creation. In other words, OCS researchers must be self-aware with an embedded intentionality, a cycle of action and reflection based on one's own lived experiences and worldviews.

Within development literature and practice, researchers and practitioners have recognized the importance of action-based, citizen-focused research for decades (Hall 1992; Hall and Tandon 2014; Tandon 2017). These researchers note that although such research methods are often deemed "illegitimate" by many mainstream academic institutions that make a strong divide between objectivity and action (Greenwood 2007), it is through engagement with community members that action research offers an opportunity to overcome the "individualist, commodity-production kind of neoliberal mindset that underlies so much social science and social policy" (Greenwood 2007: 215).

Within OCSDNet, one project that exemplifies the right-to-research approach is the Lebanese-based "Local Conservation and Development with OCS" team (see Chapter 5). The intention of this research was to engage with volunteers from rural villages on the question of water quality management in local wells. Over the course of the research, the team collaborated with local residents (who, in this case, were predominantly women) to map existing water-quality problem areas and to train local residents on how to sample and test for water quality issues. The most important part of this project was not necessarily the scientific data to emerge from the water-quality findings, but rather the increased sense of collective agency and local knowledge that was generated among local residents regarding the status of their own water supplies. In some instances, participating villagers used the findings from the research to make claims against local government to address some of their more pertinent water-quality issues.

As the example above highlights, facilitating opportunities for often-marginalized actors to be actively engaged in processes of designing, implementing, and communicating research processes has the potential to generate important positive outcomes for key development challenges. Moreover, personal agency and an understanding of local power structures are key factors for facilitating a space where citizens can actively contribute toward an inclusive OCS.

Principle 5: Equitable Collaboration

> Equitable, horizontal interaction and collaboration between formal and informal knowledge communities. We emphasize collaboration and co-creation as means for community-devised solutions and social innovation.

Principle 5 of the OCS Manifesto stresses the importance of equitable collaboration among and between heterogeneous epistemic communities in order to achieve sustainable development objectives. In this regard, it is not sufficient to merely bring people together to work. Among our key findings as a research coordination team was that equitable, long-term collaboration and co-creation of knowledge require that all members have an equitable role in shaping the nature, context, and structures for collaboration. This requires meaningful discussions around power and positionality with the intention of developing relationships and trust between co-researchers.

Within OCSDNet, one of the projects that has exemplified this principle (and advocated for its inclusion in the Manifesto) is the project from Kyrgyzstan: "Kyrgyz Mountains Environmental Education and Citizen Science" (see Chapter 12). This is an environmental education project with the aim of generating locally relevant environmental data using a citizen science approach, involving students and teachers in the pilot schools of Naryn, Kyrgyzstan.

The team stressed that in order to build relationships of co-creation and collaboration, "a culture of sharing needs to be nurtured" to showcase the value and benefits of open and collaborative science and change public perception around who is considered a legitimate scientist. The project also drew attention to the possibility that by redistributing control from "scientists" back to local knowledge holders and producers, there is increased potential to create locally relevant knowledge that responds to social demands.

Another important realization within OCSDNet has been the recognition of the challenge of collaboration within the network itself, given the highly multi-disciplinary and multilingual nature of research team, from a variety of Southern and Northern contexts. While some teams come from more "natural science" backgrounds, others are more aligned to a variety of social sciences—including law, education, and social studies of technology. Adding to the complexity was the most important, but difficult, task of ensuring equitable collaboration between Northern and Southern partners. This is difficult because Northern partners are often located in well-resourced institutions with past experiences in grant applications and funding management, putting them in a position of power relative to their Southern partners.

While the research available on long-term collaboration between Northern and Southern research institutions is limited, Brown and Gaventa (2008) suggest four core opportunities for establishing research environments grounded in inclusive and equitable collaboration. These components include: "(1) the articulation of shared values and purposes, (2) the development of relationships and trust among network members, (3) the creation of a network architecture of tasks, structures, cultural expectations, and organizational resources that shape its activities, and (4) the distribution of formal and informal power within the network." Adding to this analysis, from a development-research funding perspective, a radical shift in donor-grantee relationships and calls for proposals may be required. At present, most funding calls are still structured in ways that favour applicants from well-resourced institutions, who are better able to respond to the bureaucratic requirements and the language of academic discourse assumed or stipulated by Northern donors.

Evidently then, equitable collaboration is by no means an easy target. Within OCSDNet, we have witnessed the emergence of conflict between interdisciplinary and North-South partnerships in various ways. For instance, one OCSDNet project based in South Africa has been working in close partnership with an American university. In this case, there were significant issues with the process of receiving ethical clearance to conduct research on openness with Indigenous communities in South Africa due to the bureaucratic research-ethics requirements of the American university. Despite attempts by both the American researcher and South African partners to facilitate a transparent and reflexive ethics process driven by the local community, the American university was adamant about having pre-approved, informed consent letters, research questionnaires, etc. In this way, the American institution asserted itself as the "standard of excellence" for research practice, while the South African partners felt that these prescribed methods were both inappropriate and counterproductive in the local context and to the specific research objectives. Fortunately, the team was able to use the opportunity to engage critically with the American university's ethics department, while working with community members themselves, to develop informed consent letters that were agreed upon by all. Thus, in merging ideas both from Northern and Southern partners, the result was an innovative, high-quality, and locally appropriate approach to ethical research collaboration (Chapter 10).

In other instances, the OCSDNet coordination team witnessed logistical and ideological struggles between Northern and Southern co-investigators. In these instances, we witnessed the dominance of the "Northern" member on the team, who was often in control of project resources and hence better positioned to steer project priorities and core decision making. In other instances, it appeared that Southern partners were recruited more as figureheads positions to fulfill the Global South partner's requirement of the call for proposals, while, in reality, Global North partners and institutions largely led the projects (Piotrowski 2014). Nonetheless, despite these instances of power inequality, some of the most successful, nuanced, and robust findings emerged from teams with well-established roles, working relationships, and trust between diverse actors—whether they be North-South, South-South, researcher-community relationships, or otherwise.

In summary, there is recognition that cross-cultural and inter-disciplinary collaboration is difficult, time consuming, and requires deep dedication by all members. On the other hand, these complex forms of collaboration are incredibly important for the development of an inclusive Open Science that values deep and diverse forms of knowledge.

Principle 6: Inclusive Infrastructures

> Tools that integrate the diverse contexts and needs of all stakeholders in their design. Inclusive infrastructures promote greater interaction between data providers and data users, and enable all the actors to produce, gather, share, collaborate, and use scientific knowledge.

Building on the importance of equitable collaboration raised in the previous section, many ICT for Development (ICT4D) advocates would suggest that the increased access to and use of new technologies by marginalized communities has the opportunity to contribute to development objectives in ways that would not have previously been possible. Similarly, from an Open Science perspective, many advocates would suggest that the open source movement has created new opportunities for diverse participation, forms of collaboration and information sharing that, by their very nature, should facilitate more inclusive scientific research (McKiernan et al. 2016).

Within OCSDNet, we recognize that technologies do indeed have an important role to play in making research and knowledge-creation processes more accessible. However, at the same time, it is imperative to think critically about the role and use of particular technologies in terms of their potential to democratize knowledge-creation processes and expand the agency and decision-making capacity of users. While some "open" technologies and tools may genuinely facilitate collaboration, transparency, and inclusivity, others may simply re-create existing power relations within virtual spaces. As Powell (2012) explains: "Despite these views of open participation structures as challenging to hegemonic forms of media, tension remains between radical re-interpretations of how knowledge or culture should be produced and the co-optation of this knowledge by institutions such as the market."

With this in mind, Principle 6 of the OCS Manifesto calls for the development and use of "inclusive infrastructures" toward the creation of a more diverse and inclusive science. With the term "infrastructures," we acknowledge not only the use of ICTs, but also the diversity of tools, methods, and structures that shape or facilitate the way that research collaboration can be designed to enable users of diverse abilities to pursue knowledge production, as well as development objectives. In the scope of research within the network, we have recognized the importance of technology and tools that are locally appropriate and which seek to acknowledge and minimize competing power relations at the levels of design, implementation, and use. In the words of Denisa Kera, the principal investigator of the "Open Science Hardware Project" (see Chapter 3): "The OSH as democratized and low-tech approaches to science is an activity, tool, and community, which 'allow(s) multiple futures for science' and enables science to happen in unusual spaces, 'in or out of the academy, in or out of the lab, in or out of commercial spaces.'"

In defining inclusive research infrastructures, it is important to distinguish between technology and tools for communication versus data collection and analysis, and the dissemination of research outputs. In the case of creating inclusive infrastructures for communication, many OCSDNet projects have stressed the importance of not over-complicating the ways in which actors communicate as part of a collaborative process. For instance, in the Brazil-based project "OCS and Community Development in Brazil" (see Chapter 13), simple technologies, such as radio programming, were used as effective tools to engage communities in discussions of Open Science, while projects in West and South Africa made use of theatre and drama in lieu of standard technologies for similar engagement purposes. In South-East Asia, the project specifically embedded design-thinking into the planning of their open-hardware workshops, using iterative methodologies to improve facilitation and engagement for each successive workshop.

These examples point to the potentially empowering experience that can emerge from a more inclusive and collective process of designing, constructing, and testing new tools and processes. In other words, through a process of critical reflection on existing tools, processes, and infrastructures, many teams have recognized the need to re-evaluate and co-design new mechanisms for learning, knowledge-creation, and collaboration. The result is not only the creation of inclusive infrastructures, but also an expanded definition of what constitutes infrastructure for creative, relative, and nuanced forms of knowledge.

Principle 7: Sustainable Development

> Improving the capacity of individuals and communities to act on their own behalf and contribute to the well-being of their communities. Meaningful local development is culturally sensitive, environmentally sustainable, and led by communities.

Recognizing the ambiguities, historical legacies, and multiple meanings around the concept of "development," network members recognized that it would be important to have at least one principle within the Manifesto that would reflect a shared understanding of the term, grounded in the context of an inclusive OCS. Of course, the concept of "sustainable development" is not a new one and can be traced back to the Brundtland Commission's 1987 groundbreaking report *Our Common Future* (UN World Commission on Environment and Development 1987). This report was the first of its kind to recognize that complex global challenges could only be solved through a holistic consideration of environmental, social, and economic factors, which are intrinsically interconnected.

Over the years, this term has been taken up by development agencies and NGOs around the world, the most prominent iteration of which is currently captured within the 2015 Sustainable Development Goals. However, despite the prominence of this discourse within the development field, sustainable development has rarely been discussed within the context of Open Science. Beyond the tripartite definition of sustainable development as a recognition of environmental, economic, and social factors for solving development challenges, OCSDNet acknowledges that the creation and/or use of local knowledge is a key prerequisite for achieving sustainable development outcomes. Moreover, there is a need for communities, local institutions, and research experts to find ways to collaborate in their pursuit of sustainable development objectives and to centre different forms of relevant knowledge and ways of knowing into these shared endeavours.

Again, this is not a new realization. Through an acknowledgement of the challenges encountered through the use of top-down development infrastructures and institutions, many researcher-practitioners have been making use of participatory tools and forms of engagement since the 1970s, which allow for a more bottom-up approach to development.[4] These tools have the opportunity to produce spaces in which knowledge can be co-created by a multiplicity of actors and applied to complex problem solving. Many prominent development theorists have thus influenced OCSDNet's definition of sustainable development. For instance, Amartya Sen's (2003) capabilities approach, in which an individual's potential for freedom is seen as a standard of well-being and development, has been highly influential in our work. Likewise, scholars who encourage the consideration of diverse knowledge pluralities have been highly relevant to our discussions, including Vandana Shiva's (1995) work that considers development research within a feminist-ecological framework, and that of Arturo Escobar (1995), who situates understandings of development in local contexts of history and society.

Within OCSDNet, the project entitled "Climate Change Adaptation in Colombia and Costa Rica" (see Chapter 4) sought to work with small community groups, through a series of participatory workshops and focus groups, to better understand local issues around climate change, and to work toward developing local solutions for addressing these issues. As part of the process, citizens were given the space to act as co-researchers as well as to facilitate opportunities for collaboration between local scientists and academics. The goal was to develop nuanced, but locally appropriate solutions to pressing challenges.

Similarly, in the project "OCS and Community Development in Brazil" (see Chapter 13), the team used the concept of "sustainable development" to guide its analysis and research agenda in the context of a multitude of diverse actors (including local communities, scientists, building developers, tourists, etc.), all with competing notions of what "development" should entail within Ubatuba's fragile coastal ecosystem. Importantly, the team raised the question of "Open Science for whom?" within its research, noting that the process of designing open and inclusive research for sustainable development may change, depending on with whom you are attempting to engage.

In sum, as any development researcher or practitioner knows, development is never an easy concept to define. While it is intrinsically grounded in the idea of growth and change, mainstream cultures of consumption and production force us to think critically about issues of climate change, biodiversity, inequality, pollution, and other pressing global challenges. Hence, OCSDNet grounds our understanding of sustainable development in small-scale, local solutions that take account of local practices of conservation, problem-solving, and resource sharing where possible. Indeed, OCS advocates for practitioners to strive to acknowledge these constraints and opportunities for pursuing long-term, sustainable, and inclusive development objectives.

Conclusion

Drawing on observations from twelve OCSDNet projects, this chapter has sought to outline seven core principles that collectively illustrate a co-created understanding of an inclusive, open, and collaborative science in development. These principles range from the importance of situating inclusive scientific research in the context of a "knowledge commons," as well as acknowledging historical power asymmetries that warrant the need for knowledge pluralities through "cognitive justice." Principle 3 draws on feminist thinking to encourage researchers to "situate" their understanding of science within highly nuanced, socio-cultural terrains that shape power structures around which science is practiced within a given context. Appadurai's "right to research" is highlighted in Principle 4, acknowledging that all human beings should have the opportunity to experiment and, hence, generate knowledge that is relevant to their own context. Principles 5 and 6, respectively, outline the importance—and challenges—of constructing equitable opportunities for collaboration, while acknowledging that researchers must intentionally seek to create "inclusive infrastructures" to avoid recreating the status quo of research inequalities. Finally, Principle 7 suggests that all of these factors should be considered in the context of pursuing "sustainable development" objectives, grounded in a holistic integration of local community knowledge, respect for the environment, and the collaboration of diverse actors.

Importantly, the principles and examples presented throughout the chapter must be considered in the larger, more mainstream context of Open Science, which, to date, has largely failed to acknowledge the power structures and knowledge inequalities that exist, thus preventing many communities from participating in knowledge-creation processes. Evidently, this lack of critical discourse has negative implications for sustainable development, as marginalized groups continue to be excluded, despite the promise and allure of Open Science and its associated technologies. Indeed, the majority of Open Science policies to date have emerged from Western institutions and tend to recreate the status quo in terms of hierarchies of colonial knowledge and ways of working (Bezuidenhout et al. 2017; Albornoz et al. 2018).

For now, this chapter has sought to outline core principles for development researchers and practitioners working in cross-cultural contexts, seeking to develop an OCS environment grounded in inclusion. Admittedly, we are still a long way from being able to construct a participatory platform that is truly inclusive, given our limited understanding of how such a new system would be governed and sustained. Likewise, our understanding of the linkages between OCS and the creation of a viable knowledge commons is still in its infancy. There is much to be learned about the relationship between local Indigenous knowledge and globalized forms of knowledge, and we know little about how principles of local commons match up with those of commons at the regional and global level (Hall et al. 2012). Nonetheless, emerging evidence from the network does indeed suggest that "openness" is best understood as a process, as social praxis (Cronin 2016; Smith and Seward 2017), and as highly situated (Bezuidenhout et al. 2016).

Ultimately, while the framework bridging "OCS" and "Development" is in its infancy, this chapter suggests that an inclusive Open Science is not a new concept. Instead, it is a reflexive exercise that seeks to bring science back to its roots. An inclusive Open Science is unafraid of acknowledging and addressing other ways of knowing. As Haraway (1988) rightly says, "science has been utopian and visionary from the start; that is one reason 'we' need it" (585). OCS has the potential to be transformative, and, as Appadurai (2006) also reminds us, it has "the capacity to aspire" and the rights to research are constitutive of Development.

Notes

1. See the full draft of the Manifesto at https://ocsdnet.org/manifesto/open-science -manifesto/.
2. The annotated bibliography and collaborative reading list is available here: https:// goo.gl/us7rj7.
3. The OECD publication entitled *Making Open Science a Reality* (2015) is one recent example.

4. See our blog post (https://ocsdnet.org/open-science-and-development-the
-importance-of-cross-disciplinary-learning/) on the importance of cross-disciplinary
learning between Open Science and Development or Dr. Rajesh Tandon and Budd
Hall's work on community-based research to develop socially relevant knowledge
in Higher Education Institutes (HEIs).

References

Albornoz, Denisse, Alejandro Posada, Angela Okune, Rebecca Hillyer, and Leslie Chan. 2017. "Co-Constructing an Open and Collaborative Manifesto
to Reclaim the Open Science Narrative." In *Expanding Perspectives on
Open Science: Communities, Cultures and Diversity in Concepts and Practices*,
edited by Leslie Chan and Fernando Loizides, 293–394. Limassol, Cyprus:
ELPUB. https://elpub.architexturez.net/doc/10-3233/978-1-61499-769-6-293.

Albornoz, Denisse, Maggie Huang, Issra Martin, Maria Mateus, Aicha Touré,
and Leslie Chan. 2018. "Framing Power: Tracing Key Discourses in
Open Science Policies." In *ELPUB 2018*, edited by Leslie Chan and
Pierre Mounier. Toronto, Canada. https://doi.org/10.4000/proceedings
.elpub.2018.23.

Appadurai, Arjun. 2006. "The Right to Research." *Globalisation, Societies and
Education* 4 (2): 167–77.

Bezuidenhout, L., B. Rappert, A. H. Kelly, and S. Leonelli. 2016. "Beyond
the Digital Divide: Towards a Situated Approach to Open Data." *Science and Public Policy* 44 (4): 464–75. https://ore.exeter.ac.uk/repository
/handle/10871/21288.

Bezuidenhout, Louise, Ann Kelly, Sabina Leonelli, and Brian Rappert. 2017.
"'$100 Is Not Much To You': Open Science and Neglected Accessibilities
for Scientific Research in Africa." *Critical Public Health* 27 (1): 39–49.
https://doi.org/10.1080/09581596.2016.1252832.

Bollier, David, and Silke Helfrich. 2014. *The Wealth of the Commons: A World
beyond Market and State.* Amherst, MA: Levellers Press. http://wealthofthe
commons.org/home.

Brown, David and John Gaventa. 2008. "Constructing Transnational Action
Research Networks: Observations and Reflections from the Case of the
Citizenship." DRC. *IDS Working Paper*: 302.

Cohn, Jeffrey. 2008. "Citizen Science: Can Volunteers Do Real Research?"
BioScience 58 (3): 192–97. doi: http://dx.doi.org/10.1641/B580303.

Cronin, Catherine. 2016. "Openness and Praxis (at #SRHE)." https://catherine
cronin.wordpress.com/2016/11/28/openness-and-praxis/.

Escobar, Arturo. 1995. *Encountering Development: The Making and Unmaking of
the Third World.* Princeton, NJ: Princeton University Press.

———. 2011. *Encountering Development: The Making and Unmaking of the Third
World.* Princeton, NJ: Princeton University Press.

Frischmann, Brett M., Michael J. Madison, and Katherine J. Strandburg, eds. 2014. *Governing Knowledge Commons*. New York: Oxford University Press.

Greenwood, Davydd and Morten Levin, eds. 2007. "An Epistemological Foundation for Action Research." In *Introduction to Action Research: Social Research for Social Change, Second Edition*, 55–85. Thousand Oaks, CA: SAGE Publications.

Grigorov, Ivo, Mikael Elbæk, Najla Rettberg, and Joy Davidson. 2015. "Winning Horizon 2020 with Open Science." *Zenodo*, January 9, 2015. https://doi.org/10.5281/zenodo.12247.

Grosfoguel, Ramón. 2007. "The Epistemic Decolonial Turn." *Cultural Studies*, 21 (2–3): 211–23. https://doi.org/10.1080/09502380601162514.

Gudynas, Eduardo. 2011. "Buen Vivir: Today's Tomorrow." *Development* 54 (4): 441–47. https://doi.org/10.1057/dev.2011.86.

Hall, Budd. 1992. "From Margins to Center? The Development and Purpose of Participatory Research." *The American Sociologist* 23(4): 15–28. https://doi.org/10.1007/BF02691928.

Hall, Budd and Rajesh Tandon. 2014. "No More Enclosures: Knowledge Democracy and Social Transformation." *openDemocracy*, August 20, 2014. https://www.opendemocracy.net/transformation/budd-hall-rajesh-tandon/no-more-enclosures-knowledge-democracy-and-social-transformat.

———. 2017. "Decolonization of Knowledge, Epistemicide, Participatory Research and Higher Education." *Research for All* 1 (1): 6–19. https://doi.org/10.18546/RFA.01.1.02.

Hall, Nina, Nimi Hoffmann, and Marius Ostrowski. 2012. "The Knowledge Commons: Research and Innovation in an Unequal World." *St Antony's International Review* 8 (1): 3-12. http://www.jstor.org/stable/26229083.

Haraway, Donna. 1988. "Situated Knowledges: The Science Question in Feminism and the Privilege of Partial Perspective." *Feminist Studies* 14 (3): 575–99.

Harding, Sandra. 1986. *The Science Question in Feminism*. Ithaca, NY: Cornell University Press.

———. 2006. *Science and Social Inequality: Feminist and Postcolonial Issues*. Urbana: University of Illinois Press.

———. 2015. *Objectivity and Diversity: Another Logic of Scientific Research*. Chicago: University of Chicago Press.

Hess, Charlotte and Elinor Ostrom. 2005. "A Framework for Analyzing the Knowledge Commons." In *Understanding Knowledge as a Commons: From Theory to Practice*, edited by Charlotte Hess and Elinor Ostrom, 41–80. Cambridge, MA: MIT Press. http://surface.syr.edu/sul/21.

Kansa, Eric. 2014. "It's the Neoliberalism, Stupid: Why Instrumentalist Arguments for Open Access, Open Data, and Open Science Are Not Enough." *LSE Impact Blog*. http://blogs.lse.ac.uk.myaccess.library.utoronto.ca/impactofsocialsciences/2014/01/27/its-the-neoliberalism-stupid-kansa/.

Leonelli, Sabina, Daniel Spichtinger, and Barbara Prainsack. 2015. "Sticks and Carrots: Encouraging Open Science at Its Source." *Geo: Geography and Environment* 2 (1): 12–16. https://doi.org/10.1002/geo2.2.

Madison, Michael J. 2014. "Commons at the Intersection of Peer Production, Citizen Science, and Big Data: Galaxy Zoo." In *Governing Knowledge Commons*, edited by Brett M. Frischmann, Michael J. Madison, and Katherine J. Strandburg. New York: Oxford University Press.

McKiernan, Erin, Philip Bourne, C. Titus Brown, Stuart Buck, Amye Kenall, Jennifer Lin, and Jeffrey Spies. 2016. "How Open Science Helps Researchers Succeed." *Elife* 5: e16800. https://doi.org/10.7554/eLife.16800.

Mies, Maria, and Vandana Shiva. 2014. *Ecofeminism.* 2nd edition. London: Zed Books.

Mignolo, Walter D. 2002. The Geopolitics of Knowledge and the Colonial Difference. *The South Atlantic Quarterly*, 101 (1): 57–96. https://muse.jhu.edu/article/30745.

———. 2011. *The Darker Side of Western Modernity: Global Futures, Decolonial Options.* Durham, NC: Duke University Press.

Monni, Salvatore, and Massimo Pallotino. 2015. "A New Agenda for International Development Cooperation: Lessons Learnt from the Buen Vivir Experience." *Development* 58 (1): 49–57.

Nosek, B. A., G. Alter, G. C. Banks, D. Borsboom, S. D. Bowman, S. J. Breckler, and T. Yarkoni. 2015. "Promoting an Open Research Culture." *Science* 348 (6242): 1422–5. https://doi.org/10.1126/science.aab2374.

Nyamnjoh, Francis. 2009. "Open Access and Open Knowledge Production Processes: Lessons from CODESRIA." *The African Journal of Information and Communication (AJIC)* 10. https://www.academia.edu/24756363/Institutional_Review_Open_Access_and_Open_Knowledge_Production_Processes_Lessons_from_Codesria.

———. 2013. "Africa, the Village Velle: From Crisis to Opportunity." *Ecquid Novi: African Journalism Studies* 34 (3): 125–40. https://doi.org/10.1080/02560054.2013.852786.

OECD. 2015. *Making Open Science a Reality.* OECD Science, Technology and Industry Policy Papers. Paris: Organisation for Economic Co-operation and Development. http://www.oecd-ilibrary.org/content/workingpaper/5jrs2f963zs1-en.

Okune, Angela, Rebecca Hillyer, Denisse Albornoz, Nanjira Sambuli, and Leslie Chan. 2016. Tackling Inequities in Global Scientific Power Structures. https://tspace.library.utoronto.ca/handle/1807/71107.

Ostrom, Elinor. 1990. *Governing the Commons: The Evolution of Institutions for Collective Action.* Cambridge: Cambridge University Press.

———. 2005. *Understanding Institutional Diversity.* Princeton, NJ: Princeton University Press.

———. 2009. "A General Framework for Analyzing Sustainability of Social-Ecological Systems." *Science*, 325 (5939): 419–22.

Piotrowski, Jan. 2014. "Power Imbalances 'Still Harming North-South Alliances.'" SciDev.Net. https://www.scidev.net/global/publishing/news/power-imbalances-still-harming-north-south-alliances.html.

Powell, Alison. 2012. "Democratizing Production Through Open Source Knowledge: From Open Software to Open Hardware." *Media, Culture and Society* 34 (6): 691–708. https://doi.org/10.1177/0163443712449497.

Santos, B. de S., ed. 2008. *Another Knowledge Is Possible: Beyond Northern Epistemologies*. London; New York: Verso.

———. 2014. *Epistemologies of the South: Justice Against Epistemicide*. New York: Routledge.

Schmidt, Birgit, Astrid Orth, Gwen Franck, Iryna Kuchma, Petr Knoth, and José Carvalho. 2016. "Stepping Up Open Science Training for European Research." *Publications* 4 (2): 1-10. https://doi.org/10.3390/publications4020016.

Schweik, Charles M., and Robert C. English. 2012. *Internet Success: A Study of Open-Source Software Commons*. Cambridge, MA: The MIT Press.

Sen, Amartya. 2003. "Development as Capability Expansion." In *Readings in Human Development*, edited by Sakiko Fukuda-Parr and A. K. Shiva Kumar. New Delhi and New York: Oxford University Press.

Shiva, Vandana. 1995. "Democratizing Biology: Reinventing Biology from a Feminist, Ecological and Third World Perspective." In *Reinventing Biology: Respect for Life and the Creation of Knowledge*, edited by Lynda Birke and Ruth Hubbard, 50–71. Bloomington: Indiana University Press.

———. 2016. *Staying Alive: Women, Ecology, and Development* (Reprint edition). Berkeley, CA: North Atlantic Books.

Silvertown, Jonathan. 2009. "A New Dawn for Citizen Science." *Trends in Ecology and Evolution* 24 (9): 467–71.

Smith, Matthew Longshore, and Ruhiya Seward. 2017. "Openness as Social Praxis." *First Monday* 22 (4). http://dx.doi.org/10.5210/fm.v22i4.7073.

Tandon, Rajesh. 2017. "Participatory Research in Asia (PRIA)." In *The Palgrave International Handbook of Action Research*, edited by Lonnie L. Rowell, Catherine D. Bruce, Joseph M. Shosh, and Margaret M. Riel, 441–53. New York: Palgrave Macmillan. http://link.springer.com/chapter/10.1057/978-1-137-40523-4_27.

Tkacz, Nathaniel. 2012. "From Open Source to Open Government: A Critique of Open Politics." *Ephemera; Leicester* 12 (4): 386–405. http://www.ephemerajournal.org/contribution/open-source-open-government-critique-open-politics-0.

Tyfield, David. 2013. "Transition to Science 2.0: 'Remoralizing' the Economy of Science." *Spontaneous Generations: A Journal for the History*

and Philosophy of Science 7 (1): 29–48. https://doi.org/10.4245/sponge.
v7i1.19664.

UN World Commission on Environment and Development. 1987. *Our Common
Future*. Oxford: Oxford University Press.

Van Overwalle, Gertruui. 2014. "Governing Genomic Data: Plea for an
'Open Commons.'" In *Governing Knowledge Commons*, edited by Brett
M. Frischmann, Michael J. Madison, and Katherine J. Strandburg. New
York: Oxford University Press.

Appendix - Questions from the Knowledge Commons Framework

Questions derived from the Knowledge Commons Framework
(Frischmann et al. 2014) used for guiding responses from the OCSD-
Net projects.

Background or context:
- What is the background context (legal, cultural, political, technical, economic, etc.) of your project?
- What is the default status of knowledge resources in this context (patented, copyrighted, open, etc.) before or during the introduction of your project?

Culture of openness:
- What is the culture of openness in your policy, social, and cultural context?
- If it already exists, what are the different social, cultural, and policy angles that have contributed to this culture and awareness of openness? If it does not, what are the barriers?

Community Members:
- Who are the members of the community managing common resources and what are their roles?
- Are there any community members who benefit from openness (women, disabled, etc.)?
- How does a culture of openness affect your project's engagement with the general public?

Resources:
- What technologies and skills are needed to create, obtain, and maintain the resources at stake?
- What technologies and skills are needed to create, obtain, and maintain a culture of openness?

Governance:
- What are the governance mechanisms (e.g., membership rules, resource contribution or extraction standards and requirements, conflict resolution mechanisms, sanctions for rule violation)?
- Who are the decision makers and how are they selected?
- What are the institutions and technological infrastructures that structure and govern decision making?

Patterns and Outcomes
- What benefits (e.g., innovations and creative output, production, sharing and dissemination of knowledge, social interactions) are delivered to members of the community?
- What costs and risks are associated with collaboration, including negative externalities?

SECTION 1

DEFINING OPEN SCIENCE
IN DEVELOPMENT

INTRODUCTION

Apiwat Ratanawaraha

The first section of this volume comprises three chapters that contribute to our understanding of Open Science as practised in the context of development. Although definitions of "Open Science" and "Development" are not addressed per se, these contributions help us explore possible analytical definitions of "Open Science in Development" by describing the associated assumptions, properties, and contexts. The authors use concrete examples of research projects and their specific socio-technical contexts to illustrate how scientific initiatives can be made more open with the expectation of positive developmental outcomes.

The projects had several characteristics in common:

- All were citizen science initiatives conducted in developing countries, adopting a bottom-up, participatory approach to project development and implementation.
- Working collaboratively with local communities, the research teams were interdisciplinary, involving natural scientists, social scientists, engineers, and designers.
- They all faced challenges and opportunities associated with designing culturally appropriate research initiatives at the local level, particularly socio-technical tensions that arose from involving people with diverse, and often opposing, perspectives about science.

Yet each chapter demonstrates that citizen participation is a necessary, albeit insufficient, condition for Open Science in development.

Chapter 3 by Huang, Kera, and Widyaningrum describes the experiences of implementing Open Science projects in which international and domestic teams of researchers engaged with local communities in making scientific instruments and tools. The authors point to "informal" aspects of scientific processes as being exploratory, artistic, and speculative, and the process of using scientific experiments as a way for people to explore, discuss, and understand what science means in various contexts. They also highlight the continuous tension between the global notion of scientific knowledge exchange and the appreciation of local roots and context of science. They find that Open Science hardware is not simply about making cheaper and more accessible scientific tools. Rather, the process functions as a "social device" that fosters "little science" communities, which combine interest in science with reflections on critical issues facing the communities.

In Chapter 4, Lorenzo, Rodriguez, and Benavides examine the motivations of participants and non-participants for engaging in a citizen science project in two model forests in rural communities of Costa Rica and Colombia. In addition to studying the incentives and motivations, the research team hoped to widen the horizons of the local groups by establishing connections with the broader landscape of the Model Forest and to provide the communities with opportunities for self-organization, including defining the problems and establishing local priorities. The authors highlight the multi-motivational nature of involvement in projects that require a high level of voluntary engagement. Building on the standard Participatory Action Research (PAR) approach to conducting a research and advocacy project, the research team emphasized the notion of "reciprocity," that is, the importance of not only taking but also giving back to the community throughout the research.

In Chapter 5, Talhouk et al. report their findings from a citizen science project for water quality testing in a Lebanese village. The authors describe the methodology and process to engage and train community members, as well as the responses and exchanges among the parties involved. Throughout the project, the researchers adopted an open information-sharing framework between the academic team and the community. They show evidence that citizen science can be used as a tool for community development and that it has the potential to build a social foundation for remediating local environmental problems.

The three projects share several basic principles of Open Science as identified in the Open Science in Development Manifesto proposed by the Open and Collaborative Science in Development Network (OCSDNet). These include the value of plurality and diversity in science; the use of frameworks, mechanisms, and tools that help correct existing imbalances in power and resources in producing and sharing knowledge; the opportunities for participation at all stages of the research process; and equitable collaboration between scientists and social actors. In addition, the three projects share two interrelated characteristics that add to the analytical definition of Open Science in development: namely, Open Science as a mechanism to improve transparency, and pragmatism as the underlying philosophy of Open Science in development.

Open Science as a Mechanism for Improving Transparency

In his 1999 book *Development as Freedom*, Amartya Sen argues that development is not simply about increasing income levels but more about an array of overlapping mechanisms that enable individuals to exercise a range of freedoms. In addition to freedom of opportunity and economic protection from extreme poverty, a fundamental condition for enhancing development as freedom is to improve and guarantee transparency in relations between the government and citizens, and among citizens themselves. In Sen's words, citizens should have "freedom to deal with one another under guarantees of disclosure and lucidity" (1999: 39). Guarantees for openness and disclosure, plus rights to information, among others, are therefore essential to development as freedom, especially in increasingly complex and pluralistic societies.

Removing existing constraints of transparency guarantees requires public discussion and deliberation. Compared to private dealings behind closed doors, public forums give citizens more opportunities to become engaged and open to one another, creating room to express and possibly accept different views and perspectives. This affirms extensive and expansive roles of civil society, specifically citizens themselves, in any public projects. Such public projects are not limited to public works that have direct impacts on the well-being and livelihoods of citizens, but also include scientific and technological endeavours that could have long-term implications for society as a whole. Particularly relevant here are scientific activities that may provide evidence to confirm or reject the underlying ideas of certain public policies and their implementation at the local level. The water quality testing and model forest projects are cases in point.

The three papers show the potential of Open Science as a way to improve transparency relations among the government, scientists, citizens, and other stakeholders, and thus a mechanism for development as freedom. Scientific activities as public projects always occur against the backdrop of a particular set of relationships between the government, firms, citizens, and other social actors. As detailed in the papers, Open Science in development necessarily involves deliberation and negotiation among various actors who are involved in the process of creating, sharing, and utilizing knowledge, regardless of scientific issues, locations, and contexts. As a result, the relationships between scientists and other social actors are redefined and made more transparent, possibly leading to better allocation of resources and public policies that support the improvement of other types of freedom.

Pragmatism and Communities of Inquiry

Another aspect shared by the three projects that define Open Science in development is pragmatism. As Charles Sanders Peirce, John Dewey, and other thinkers in the school of pragmatic philosophy of science contend, science is best viewed in terms of its practical uses and outcomes. Pragmatists do not merely discuss and debate ideas but act on their practical application by testing them in actual events and projects. Based on this definition, development is necessarily pragmatic in that the improvement in well-being and livelihoods of people has to be tangible and real, if not always measurable.

To pragmatists, creating a "community of inquiry" is necessary in the scientific process of knowledge creation and sharing. A community of inquiry is formed when people engage in a collaborative process of conceptual or empirical inquiry to identify shared problems and to develop agreeable solutions. For such a community to function well, three basic conditions are required: free inquiry, free association, and free communication (Dewey 1939/1998: 342). To Dewey, communication is particularly important not just because it is a means of transferring information and ideas, but also because it serves as a process of "world-making"—that is, "the construction of a universe of shared meanings that brings about an enhancement of the immediate quality of experience for those who participate in it" (Neubert 2009: 23).

In line with the pragmatic definition of science, all of the projects in this section deployed some form of Participatory Action Research (PAR), which aims to understand the world while changing it for the better. PAR as a research methodology emphasizes participation in actual events that are the target of inquiry, as well as constant communication among participants. Researchers in all three projects were not merely independent, objective, and disinterested observers; they were actively engaged in the process of learning, experimentation, and communication with other scientific and social actors. To that end, they formed communities of inquiry that addressed specific issues and challenges while developing context-specific solutions. The exchanges of ideas and information among the researchers and communities were in line with Dewey's observation about the dual roles of communication in a community of inquiry, as mentioned above.

The three projects also emphasized the roles of local stakeholders in establishing the legitimacy of information and knowledge. Through deliberation and negotiation, the stakeholders in each project somehow and somewhat reached agreement that helped move the process of creating, sharing, and using knowledge forward. The legitimacy achieved by such inter-subjective agreement diverges from that of traditional scientific experiments in closed laboratories. In the Cartesian model of fixed and unchanging reality, legitimacy comes from objective assessment, generalizability, and reliability. The pragmatic approach, on the other hand, focuses on knowledge that is socially embedded and derives legitimacy from agreement between people who are involved in the process. This conceptual stance is evident throughout the three projects.

In conclusion, the three chapters in this section illustrate that Open Science in development can be analytically defined only with specific details about the processes, contexts, and outcomes of the scientific projects in question. Perhaps the people who can define it best are those directly engaged in the actual activities on the ground—not those of us who are writing about them thousands of miles away.

References

Dewey, John. 1939/1998. "Creative Democracy: The Task before Us." In *The Essential Dewey, Volume 1: Pragmatism, Education, Democracy*, edited by Larry A. Hickman and Thomas M. Alexander. Bloomington: Indiana University Press.

Neubert, Stefan. 2009. "Pragmatism: Diversity of Subjects in Dewey's Philosophy and the Present Dewey Scholarship." In *John Dewey Between Pragmatism and Constructivism*, edited by Larry A. Hickman, Stefan Neubert, and Kersten Reich, 19–38. New York: Fordham University Press.

Sen, Amartya. 1999. *Development as Freedom*. New York: Random House.

CHAPTER 3

Open Science Hardware (OSH) for Development: Transnational Networks and Local Tinkering in Southeast Asia

Denisa Kera, Hermes Huang,
Irene Agrivine, and Tommy Surya

This chapter is dedicated to Imot, a dear friend, artist, and organizer who will be deeply missed in our future Open Science and maker adventures.

Abstract

The two-year OCSDNet project on the so-called making, hacking, and tinkering practices in science (Open Science Hardware–OSH) revealed a tension between globalized notions of knowledge production and exchange (OSH as transnational infrastructure) and local practices (tinkering with OSH and science). This challenges the usual descriptions of citizen science offering a model for Open Science in the Global South. The idiosyncratic, creative, and exploratory (mis)uses of OSH instruments in various workshops critically reflect upon the agenda and institutions of science and technology for the Global South. We refer to these practices as "little science" and claim that they contrast not only with the goals of professional science, or "big science," serving industry needs, but also traditionally defined citizen science, which involves amateurs and citizens helping professional science to achieve its goals. OSH instruments are tools supporting situated and tacit knowledge and explorations closer to cooking and crafts rather than professional laboratory work. The material engagement with making OSH instruments is part of community development efforts, which created spaces for experiencing, playing with, and discussing the relations between science and community, translational exchanges of knowledge with local tinkering, and even speculations about regional and South to South networks. OSH is not simply a cheaper and more accessible infrastructure, but a "social device" supporting diverse communities around "little science" projects to define their own visions and uses of science. Through prototyping and dialogue, the participants co-create their agenda for science, define opportunities for situated learning in diverse contexts, and reflect upon the goals and futures of science in physical and digital spaces.

Introduction

"Open Science Hardware" (OSH) employs open source principles, licences, and non-digital and digital (3D printers, laser cutters, and other tools operated by computers) fabrication technologies to design and build science instruments. All open source technologies, such as the Linux computer operating system or the Arduino microcontroller platform, offer an alternative to the established, patent systems of innovation and R&D that preserve the status quo in various industries (Gortych 2014; Bessen and Meurer 2008; Haunss 2013), and the technological and science divides (Lee 2015; Maclurcan and Radywyl 2012).

"Open" means simply leveraging transnational collaborations and networks to improve the design of any tool and instrument. In the case of OSH, this includes not only software and hardware developers, but also scientists, whose aim is to improve the accessibility, quality, and affordability of various science instruments. The resulting OSH enables independent research, but also citizen science cooperation with professional scientists (Gura 2013; Sobkowicz 2011; Franzoni and Sauermann 2014) and unexpected uses of instruments, closer to community development, which we observed in Southeast Asia.

Throughout 2015 and 2016, we conducted seven OSH workshops in Indonesia, Thailand, and Nepal, which included one ten-day workshop in Yogyakarta, five workshops in Bangkok over the course of one to two days, and one ten-day workshop in Kathmandu. The goal of the workshops was to understand how OSH instruments engage local communities in research and education and to assess the potential of citizen science as a model for Open Science efforts in the Global South. The workshops' programs were open to existing creative (mis)uses of science instruments that we observed in the hosting organizations (for example, practices and intersections of art, design, and craft in Yogyakarta, Indonesia), but also to new practices we initiated in Thailand and Nepal.

OSH as an example of open source infrastructure creates opportunities for the Global South to join and even lead open source projects (De' et al. 2015; Birtchnell and Hoyle 2014). It also exposes the structural issues behind distribution of wealth and influence that prevent rapid development of open technology, science, and innovation in the Global South (Takhteyev 2012). The study of OSH in South and Southeast Asia explored the tension between the transnational aspects of OSH activities and the local uses and practices of making, crafting, and tinkering. The transnational aspects behind Open Science infrastructure and the open source movement often support the "big science" goals of creating cheaper tools for doing science in the Global South (discussed in this chapter under the heading OSH Repositories). We were surprised to see that participants in our workshops used OSH instruments to imagine a different type of science than the one we consider "standard" and "professional." We decided to refer to it as "little science" (Egghe 1994; Carillo and Papagni 2014; Price 1986) supporting tacit knowledge and direct participation, which balance transnational OSH goals with the open and hybrid goals of local tinkering.

The situated, participatory, and tacit knowledge gained through OSH practices defines science as a search for an alternative to the projects of "professional" and "big" science. OSH supports "little science" as a model for the Global South. It is a science without links to any large industrial and military interests and university ranking systems based on closed journals (Moore et al. 2011; Forero-Pineda 2006; Livingston 1976; Dickson 1988). Its heterogeneous connections and collaborations embrace the local culture and involve communities by supporting their everyday life practices, but also Indigenous knowledge and experiences (Sillitoe 2007). We claim that OSH emphasizes tacit knowledge, which extends the meaning of the "right to science" to direct engagement with how science is "produced," and empowering communities to define their own future of science. In what follows, we will discuss these three aspects of OSH (tacit knowledge, empowerment and the "right to science," transnational networks) to define the opportunities, as well as tensions behind the concept of "little science" (rather than Open Science) for the Global South.

OSH Enabling Tacit Knowledge and Tinkering in Science

OSH's ability to return science to its material "roots" through instrument building, including crafts and repair culture, supports tacit and situated knowledge (Busch and Richards 2004; Gascoigne and Thornton 2013). It brings science closer to the everyday life and the cultural context of a given community and place, which we experienced first-hand in our workshops in different locations. OSH simply enables a science that is free to explore different relations to society, and also culture, rather than to insist on its demarcation from art, religion, humanities, and social sciences (Nowotny et al. 2001).

Tinkering (Nutch 1996; Griffiths 2013; Bock and Goode 2007) is an approach to science that emphasizes the tacit explorations and convergences of social, creative, and technical experiences and ideas common in open source hardware projects (Mellis and Buechley 2011). Rather than only reproducing existing practices of science in new locations or bridging the "divides" by making Open Science in the Global South part of the professional and internationally recognized networks, the OSH tinkering challenges how science relates to society, culture, and industry.

The tacit knowledge brings projects that empower the local communities to imagine and practise their own ideas about the future of technology and science in their communities and to even question the OSH and DIY as a model serving their needs (Kaiying and Lindtner 2016). The emphasis on the local, tacit DIY practices using the transnational networks of open source technologies simply enables alternative and plural understandings of the science in the Global South.

This can have a form of a more socially and community-oriented tinkering with science, as we witnessed in Indonesia, where creative forms of "hanging out" (*gotong royong*) bring new experiences of science in everyday life through food and art. It can also take a form of a very ambitious effort of building hardware kits to transform global education in science and technology and send private microsatellites to space, which we witnessed in Nepal. It can also remain ambiguous about the relation between the local and transnational goals of OSH tinkering, which we witnessed in Thailand, where some projects tried to diffuse existing educational technology ("Littlebits") or support university-industry research and collaborations (DIY electroencephalogram) (https://storify.com/teon_io/diyscithai) while others created original solutions for small enterprises in organic agriculture through experimentation with DIYBio research around certification, heavy metal sensors, livestock, and pest control.

Rather than supporting large-scale scientific and technological projects with national and international significance, the tacit involvement with science over OSH gives new, more participatory meaning to Article 27 of the Universal Declaration of Human Rights on the "right to science." The OSH and its localized and unique forms of tinkering are examples of "little science" (Price 1986) in the Global South, which can critically question the industrial and military goals of "big science" (Carillo and Papagni 2014). While in the next chapter on the "right to science" we will describe the aspirational goals of these transnational OSH networks on the example of GOSH (Gathering for Open Science Hardware), the rest of the articles confront these aspirations with the actual documents (OSH repositories) and practices (local tinkering) we encountered on the ground in our project.

OSH Extending the Meaning of the "Right to Science"

The emphasis on open hardware in science, together with calls for Open Science, open data, and open access to journals (Neylon and Wu 2009), extends the meaning of Article 27 of the Universal Declaration of Human Rights (and the related Resolution 4.52 from 1952, entitled "Study of the 'Right to Participate in Cultural Life' [...]," Section 4/1) from indirect to direct forms of participation.

Article 27 defines our "right to science" as follows: "Everyone has the right freely to participate in the cultural life of the community, to enjoy the arts and to share in scientific advancement and its benefits." The act of building a DIY microscope or other, previously inaccessible tools, and collecting and sharing data without intermediaries extends the meaning from only "enjoying the benefits of arts and scientific advancement" created by someone else to directly participating in the creation and definition of scientific research, its goals, and even reflecting directly upon its policy.

The democratization of science in such direct participation in the processes of building and using science instruments is gaining momentum through various citizen science movements—makerspaces and hackerspaces activities (Kera 2014)—but also through the creation of a new international network and its annual "Gathering(s) for Open Science Hardware" (GOSH)[1] in different parts of the world.

The high hopes behind the OSH movement are well summarized in the GOSH Manifesto formulated in the first meeting in 2016 held at CERN in Geneva. OSH developers together with "users" of such tools (citizen scientists, scientists, designers, researchers, artists, etc.) defined the goal of OSH as infrastructure, which increases "the diversity of people with tools to perform research for knowledge discovery and for applications such as education, technological innovation, and civic action" ("Global Open Science Hardware (GOSH) Manifesto" 2016). This aspect of the OSH as infrastructure for the Global South was also emphasized on several parts of the document: "Open science hardware has the clear potential to be useful in low-resource settings including labs in the Global South" (GOSH Manifesto 2016). It even mentioned digital fabrication and lowering the price of setting up microfabrication as essential in this respect:

> [T]he decentralized production chain enabled by modern digital fabrication methods potentially opens up new markets and business models, for example manufacturing of scientific instruments in countries that experience difficulties importing specialized equipment. (GOSH Manifesto 2016)

While OSH as an infrastructure for research and education in the Global South was clearly stated in the manifesto, our team insisted on acknowledging the more communal, social and creative "(mis)uses" of instruments outside of research and education, which played an essential role in how science was performed and reflected in the workshops which we organized. Denisa Kera, one of our collaborators (and a co-author), gave a keynote at the 2016 GOSH meeting addressing these more exploratory and emancipatory uses of instruments, which go beyond research and education to enable more personal and localized experiences with science and technology.

The part of the manifesto that resonates with this agenda that we brought to GOSH relates to the statements on how OSH supports the diversity of people taking part in the research in terms of their background and interests, but also their countries of origin: "Indigenous/Non-scientist peoples can make research in their native language and adapted to their local context," and that it "aims to make cultural change so these opportunities are intergenerational." The OSH in this sense extends the right to science to active and plural participation in defining the goals and practices of science by individuals and groups beyond the insistence on "Science Technology Engineering Math" (or STEM) or the "West" as the only model and goal.

OSH Repositories and Documentation Negotiating the Tensions Between Infrastructure and Community

To understand the meaning of these more exploratory and "open" (mis) uses of OSH that are central for "little science," we can look at the example of existing OSH repositories. The OSH repositories of projects document the outcomes of tinkering, but they often fail to document with similar clarity the actual uses of these tools in concrete projects, workshops, and communities. The rare exception is the Public Lab Project (https://publiclab.org/wiki), which pays equal attention to OSH as infrastructure and community engagement, in which we see that OSH works best in places with active nongovernmental groups that have a clear agenda and already work with policy makers. The work on the infrastructure simply gets more attention because it is easier to capture, but this misses the rich and difficult-to-categorize local uses and tinkering. Our goal was to capture and see how these artistic and exploratory uses of OSH, which go beyond education or research, connect science with community and offer a different view of what is science.

Most OSH repositories concentrate on the infrastructure as a technical issue, and they try to standardize the processes of attribution and sharing of the blueprints and designs. In the case of hardware that means mainly Computer Aided Design files (CADs), but also the various "recipes," protocols or instructions on how to make things and use them for science. The main goal of OSH repositories is to enable anyone to study, modify, create, and distribute the designs of the physical objects (from robots to agricultural machinery or science instruments). The "openness" in this sense depends simply on "readily-available components and materials, standard processes, open infrastructure, unrestricted content, and open-source design tools to maximize the ability of individuals to make and use hardware" (Open Source Hardware Association 2017).

Whether this "openness" of infrastructure brings new understandings of inclusion, participation, and collaboration in science and its interaction with various communities was one of the interesting questions that came out in our first workshop in Indonesia. In the workshop, we tried to impose the co-creation of a set of basic tools to do science anywhere in the world, and we believed that the main outcome should be a wiki-style repository, which "gives people the freedom to control their technology while sharing knowledge and encouraging commerce through the open exchange of designs" (Open Source Hardware Association 2017). These basic tools in our cases had to include microscopes, turbidity meters, centrifuges, etc.

In this first workshop in Indonesia, participants spent more time reflecting and discussing the actual uses of instruments in possible scenarios rather than looking at their designs. Participants also spent a disproportionate amount of time learning how to record their experiences into the available online repository system employed, the "wiki" (http://oshw.honf.org). We also realized, in a multicultural environment, the difficulty of producing multi-language documentation in a central system. This was especially true in the case of Burmese language at the time of the workshop, when a standard Burmese font set was not yet available or, otherwise, exceedingly difficult to use. We realized that the definition of OSH "openness" conveys some implicit expectations that instruments will serve the goals of scientific research and technology innovation as we know it from the "West." This felt unrealistic, but also reductionist, since it ignored what we noticed as a more valuable effect: the tacit experiences with OSH that enabled participants to imagine different uses and visions of science for development. OSH repositories rarely, if ever, document and discuss how these tools can enable people to question the role of science in their community or enable a different view of the future, which emerged as a central insight, point of discussion, and output from the participants in our workshops.

Repositories, such as Joshua Pearce's OSAT of Thingiverse collection, Bryan Bishop projects, Open Source Ecology and Open Manufacturing initiative, and TEKLA lab, support efforts in education, research, and entrepreneurship, which are global and transnational. The more idiosyncratic and exploratory uses of OSH, which enable people to see a different "future of science" and connect it with new domains of practice and knowledge, as witnessed in our workshops, proved to be difficult to document. They often included personal narratives and documentations of events, such as workshops, exhibitions, or performances on social media bringing rich social, political, cultural, and even aesthetic contexts.

The difference between the emphasis on OSH as an infrastructure and as a community is visible when we compare the flagship academic OSH project with that of grassroots or community OSH projects. An example is the Open Source Appropriate Technology (OSAT)[2] by Joshua Pearce and his Open Sustainability Lab at the Michigan Technological University with the Hackteria's Generic Lab Equipment[3] repository. Another example is the Hackteria repository, which presents citizen science projects and design in rich context, including the reasons and settings in which the work on the instrument happens, as developed in various settings in Indonesia, India, Switzerland, Taiwan, and more.

The OSAT's wiki offers a blueprint for more efficient science infrastructure in a context of a larger project trying to open source all technological infrastructure in science as a way of creating a new business model for universities anywhere in the world. It is similar to Joshua Pearce's Thingiverse collection,[4] which summarizes all the points of his numerous articles and books on how OSH makes science more productive (Pearce 2012, 2014, 2015). The alternative, non-academic Hackteria repository and the similar GynePUNK repository of tools (http://gynepunk.tumblr.com/) sometimes describe the same instruments, but their functions remain open for artistic and exploratory uses and linked to various social and political discussions and events.

The "alternative" repositories question the goals of "big science" and the current status quo as the only possible model to follow. Thus the GynePUNK collective—self-described anarchofeminists and transhackfeminists at the Pechblenda biolab in Calafou, Spain—often collaborate with collectives in Colombia, as well as Indonesia, to address gender issues through hardware design in their repository. Their work on the instruments is founded in a broad range of considerations from art and science to politics and philosophy, which are expressed both in the workshops and in their documentation.

The types of OSH projects and uses in our workshops show a similar diversity of goals and ideas about science, which are captured through photos, Facebook posts, or videos with performative rather than descriptive value. This documentation[5] is often motivational, mobilizing other members to try at their events, rather than descriptive and didactic. This plurality of forms of capturing and describing OSH shows a tension between the global aspirations of the OSH, which wants to become a model for more efficient and independent science, and the local practices and tinkering with instruments, which bring science closer to everyday life activities (food, social interaction) and even question the role of science in their community or try to practise "little science" (Price 1986).

OSH Supporting "Little Science" in Indonesia, Thailand, and Nepal

Our OSH workshops in Indonesia (2015), Thailand (2016), and Nepal (2016) provided further examples of this tension between the OSH seen as a transnational infrastructure for doing science and the localized and tacit tinkering with OSH. OSH in the Global South proves to be more than just technical infrastructure to resolve science and technology divides. The tinkering practices with various OSH tools (microscopes, sensors, etc.) created opportunities to question the role of science in the communities. They enabled participants to imagine different "futures" of science, such as a space program education in Nepal (Figure 3.1) or brain-computer interface research in Thailand (Figure 3.2), but also science as more integrated in everyday activities—a creative form of "hanging out" and social interaction in Indonesia.

Figure 3.1. K_Space Workshop at Karkhana in Kathmandu, Nepal.

Photo Credit: Karkhana.

Figure 3.2. Building DIY Microscopes in Bangkok, Thailand.

Photo credit: Hermes Huang.

OSH supporting local tinkering, crafts, and various forms of tacit knowledge and everyday engagements with science rather than a transnational goal of developing new open hardware tools is a form of "little science" (Price 1986) rather than "citizen science." Instead of supporting amateurs in helping "big science" projects by collecting data (Franzoni and Sauermann 2014; Fienberg et al. 2011; Lewenstein 2004; Smith et al. 2010) or doing other activities with their instruments, typical of participatory monitoring projects, the OSH in our workshops supported informal and social daily interactions surrounding the scientific practices.

This "little science" (Price 1986; Lievrouw 2010; Borgman et al. 2007) model embraces tacit, culturally embedded, and situated knowledge, in which participants build tools and other equipment (Figure 3.3) while discussing and integrating these implements in their communities. The workshops and activities had very open goals and horizontal structure. The participants were mostly young students, artists, designers, entrepreneurs, and enthusiasts coming from various institutions (universities, schools, village councils, small companies, museums, and galleries). They had equal stakes in the projects, which were temporary (usually workshops and/or exhibitions) rather than long-term–oriented.

Figure 3.3. Building a DIY Turbidity Meter in Yogyakarta, Indonesia.

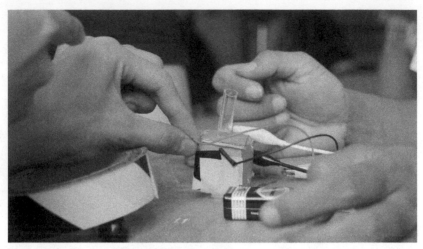

Photo credit: House of Natural Fiber Foundation.

Even when the workshops involved more ambitious agendas, such as transforming education (Nepal) or transforming agriculture to involve DIY Internet of Things (IoT) solutions (Thailand) or experimenting with brain-computer interfaces (Thailand), most of the activities revolved around discussing and learning from each other rather than achieving some clearly defined objective. For example, all HONF (House of Natural Fiber, our main collaborator) projects in Indonesia engage the general public in science as a "way of life," which means tinkering with friends and building tools for creative engagements with science, including design, crafts, and art. Between 2015 and 2016, HONF organized over fifteen workshops, which were not always officially included in the OCSDNET project but that related closely to the workshop from 2015. The topics ranged from fruit and soybean fermentation or building DIY radio antennas for astronomical data and signals (Figure 3.4), to sonification and visualization of photosynthesis processes in plants, creating artificial skin from bacteria-producing cellulose from soy waste, and making a weather-based mood lamp (using a light-emitting diode [LED], an Arduino microcontroller, and a connection to a weather Application Programming Interface [API]).

Figure 3.4. Building DIY Antennas in Nepal.

Photo credit: Karkhana.

These everyday activities and contexts (food, lamp, experience of night sky) define the "little science" model behind the OSH workshop as more suitable for empowering the Global South. Rather than citizen science research on astronomy, food science, or horticulture, these workshops support social interactions and discussions about science while participants share technical knowhow and experiences with science protocols. We could see this during the "Laboratorium Jalan-Jalan" (Mobile Laboratory) workshop, which happened in Solo and Surabaya in September 2016, or in the "Geek Diplomacy" workshops (May and September 2016).

The goal of these workshops was to establish and maintain local networks between geeks from various cities and connect the locals with visitors outside of Indonesia rather than to start a citizen science collection of data. The "Geek Diplomacy" workshop involved an artist community named Garis Cakrawala Visual Art Company in Ruang Seni (Art Space) DAYA JOEANG, in Solo, while "Laboratorium Jalan-Jalan" (mobile laboratory) involved a UK-based visiting designer and geek, Vicky Gerrard, and our co-author and the Project PI, Irene Agrivene, who worked with Ruang Atas Community and its Muara Market space in Solo.

The "Geek Diplomacy" workshops taught basics of microbiology (sterilization, fermentation) through the protocol for making a pineapple wine. They also explored the basic Raspberry-Pi[6] open-source computers and Arduino[7] microcontrollers to program sensors, potentiometers and LEDs for DIY water conductivity meters. By creating pineapple wine in a collaborative environment, participants were able to connect food with science through a social discussion of fermentation. In Indonesia, these gatherings also touched upon taboos and contemporary tensions surrounding alcohol drinking in Islam, Indigenous practices of wine-making, and economic issues around the high taxation of alcohol. In fact, these workshops were conceived after multiple citizen poisonings directly related to in-home alcohol production and organized as a response to an emerging public health issue. The workshops in this sense served as community-building events, which can face and connect tensions and taboos. Similarly, the Arduino/Raspberry-Pi workshops enabled the participants in Indonesia to discuss and test their water quality, which is another contested issue.

None of these projects has led to long-term citizen science involvement, or trying to resolve scientific or policy issues. The participants did not even expect measurable and real-world results, but rather saw the events as an opportunity to meet people with similar interests and concerns about the community, and to initiate discussions and experience various strategies of tackling these issues. While there is always a potential to turn the OSH experiences and knowledge gained through such informal interactions into a "real" citizen science project, this path was never taken during the two years of our observations and interventions.

The workshops, however, did serve to expand the possibility in the participants' daily lives. They preferred to make science practices part of their everyday lives, such as eating, having fun with friends, traveling, exploring, making drinks, etc., rather than to gain a scientific perspective on their everyday life or to use science as a way of resolving issues in their communities. A very important aspect of these "little science" interventions with OSH was the idea that science can happen anywhere. The fermentation in the "Geek Diplomacy" workshops happened in a parking lot, and the Arduino and Raspberry Pi workshops were held in a private home while having a meal.

In this sense, we define the Indonesian (and to a large extent Thai and Nepali) concepts of "openness" in OSH projects as a tacit practice deeply embedded in the everyday and social lives of the community, rather than something transgressing and transforming communities to connect them with some professional "big science." The goal of the OSH workshops was to enhance everyday practices and lives through making tools, which bring new perspectives on the future. Most citizen science projects go the opposite way; they turn the local context into data useful for some big science or large-scale programs on social innovation and innovative society, Research and Development (R&D), etc. The "openness" in the OSH is about material and tacit engagements with science, which serve personal and communal needs rather than supporting any institutional or national goals.

OSH's Transnational Networks in the Global South

OSH as a tool supporting "little science" in the Global South means negotiating the local tinkering practices (fermentation, crafts) and tools with the transnational OSH design and practices. The importance of the transnational networks and individuals supporting the OSH practices in Indonesia was visible in our initial observations related to the OCSDNET project and captured in research (Huang 2015) and another paper (Kera 2015). This result was repeated in the workshops, which also involved international participants, expats, as well as locals who studied or worked abroad.

The 2016 "Laboratorium Jalan-Jalan" (*jalan* means "street," so it is a "Street Lab") project co-organized by HONF and Vicky Gerrard represents the cosmopolitan nature of the "geekdom" involved in OSH and similar projects in the Global South very well. Vicky Gerrard focuses on co-designing "products, spaces, systems and experiences which support more inclusive approaches to social change through design"[8] especially in the health technology domain, in which she worked in India, Singapore, Myanmar, Philippines, and Cambodia. In Indonesia, she set up a mobile lab and a design studio in a 1972 Volkswagen Camper Van called Cobanana,[9] which means "trying" and tinkering (*coba*) in Bahasa. The workshops, which she co-organized with HONF, included not only the basic fermentation protocol, but also making an Atari Console as documented on Vicki's weblog. The local tinkering with science actually works by using an everyday object, such as a car, to do something new and unexpected with science, which suddenly becomes part of the everyday experiences.

The importance of the transnational networks in OSH was also visible in the planning of the Thailand 2016 workshops. The original site was supposed to be Makerspace in Chiang Mai, Thailand, which is similar to the site of our first workshop in Yogyakarta. Both spaces are part of a transnational movement of Makerspaces and FabLabs (Moilanen 2012), which claim to be creating an infrastructure for innovation in the Global South, but also solutions that are useful for the local communities (Blikstein 2013).

While the Indonesian space, HONFabLab, still shares the more complex HONF history (Kera 2012) and partially resists the start-up focus, Makerspace Thailand is a space that we can easily imagine anywhere in the EU or Silicon Valley. It is a 250-square-metre, fully equipped space with digital and non-digital fabrication tools. It has a strategic partnership with a local coffee shop, a co-working space, and the local creative hub supported by the government known as Thailand Creative and Design Center, which has other branches in Bangkok and a forthcoming branch in Khon Kaen. The founder, Nati Sang, is also a global citizen, a Thai-American from California who lived in Thailand for over ten years.

The original discussions with Makerspace Thailand were centred on creating a science laboratory infrastructure to add to the other fabrication tools in the centre, which would have just repeated the workshop model that failed in Indonesia. Makerspace Thailand had already started looking at the development of research around air quality, which is an annual issue related to swidden farming—or practising swidden agriculture, also known as shifting cultivation—in Northern Thailand driven by large agricultural conglomerates. We were interested in this because it could connect science and design with complex issues (government, food and agriculture, local practice, and industry), but we decided to change our focus based on the experience in Indonesia.

While reflecting upon the Indonesian workshops between October 2015 and January 2016, we realized that we would repeat the same mistake made in Indonesia if we insisted on the infrastructure and a set of tools rather than the more complex issue of what it actually means to do science in the Global South and how OSH tools can serve communities. We decided to try a different model, which started with defining the community interests in Open Science before pushing the agenda of infrastructure and OSH. To do this, we used the rich experience of Hermes Huang, a collaborator and co-PI in the project, who was running design thinking training and workshops through the organization DSIL Global.

The decentralized format of soliciting, rather than offering workshops, and mapping the interests started with a call for local maker, hacker, and university groups to propose and run their own workshops around Open Science and hardware. This produced some interesting data on how the local Thai communities understand Open Science, which was different from the Indonesian context and from what we later experienced in Nepal. The call was released in February 2016 to individuals living in Thailand to run "short workshops focused on science and designing hardware for science, where science is broadly defined and explored through research, art, design, education, engineering, and more."[10] Since the deadline was very short, only two applications for workshops in Bangkok, Thailand, were supported. The team also organized a workshop for the facilitators of these workshops and for people generally interested in the OSH projects in Thailand in March 2016.

The first workshop for the facilitators was organized at the Fab-Cafe/FabLab Bangkok by Denisa Kera, a member of our team, and Yair Reshef, an open hardware developer based in Singapore at that time. The goal was to meet the makers and hackers from Thailand who were already involved in OSH and plan to offer them a "library" of microcontrollers or "next generation" IoT tools with Wi-Fi and GSM capabilities. This workshop showed that most of the individuals involved in the Thai maker and hacker projects had very cosmopolitan origins, career paths, or educational backgrounds connecting Thailand with the US and Canada, both in their personal and professional lives.

In the Thai context, both the formal and informal institutions of education and research (universities, makerspaces, hackerspaces, etc.) had more connections to the "West" than to Indonesia, which led to a very different dynamic. While the Thai citizen science geeks and makers preserved a strong focus on local issues, such as health and agriculture, their use of OSH was close to any western organization. It was also surprising to notice that open hardware is actually produced in Thailand, even if the local user base is still small in comparison to places such as Shenzhen, China. The participants in the first Thai workshop were already involved in ambitious projects, such a creating IoT infrastructure for collecting sensor data from crops on an organic farm, but also urban farming. They were also exploring 3D-printed designs for health in the Thai hospitals, etc.

While the cosmopolitan and international nature of OSH efforts in Indonesia often involved organizations and individuals from abroad, who had come for a visit or to work on a project in the case of Thailand and partially also Nepal. They were mostly expats who relocated for family or work reasons or local returnees who studied abroad, mainly in the US. In this respect, the situation in Thailand could have become similar to that in Shenzhen, China, where the manufacturing industry supports innovation with a distinct transnational nature (Lindtner et al. 2015).

The second workshop was organized by the FabCafe staff for children and parents on how to use a very well-known (and expensive) Little Bits platform to create drawings of robots and learn concepts from math. The third workshop was then organized by a Thai and American pair of scholars trying open brain data tools, which, as workshop organizers Teon Brooks and Piya Kerdlap stated, served to stimulate "greater interest among young Thai people in exploring science and to build a strong sense of scientific curiosity for carrying out experiments independently." The participants were "seeking to raise awareness among young people in Thailand about the educational resources available online and low-cost methods for building equipment to carry out scientific experiments." According to the scholars, the last workshop achieved the goal through the following activities:

1. Teach students about how the human brain works in a fun and hands-on environment through simple experiments and demonstrations;
2. Provide young students with greater access to low-cost equipment and data resources to carry out experiments and conduct research independently; and
3. Empower students to make their own experiments and facilitate their own education in the sciences.

In all these workshops, the organizers were individuals with complex, transnational experiences. The third workshop gathered a demographic that would typically not meet to learn together in a non-institutional environment in the Thai context—undergraduate students and professors. This was described by the workshop organizers as a strength of the OSH approach and as an "opportunity to disrupt this conventional standard and have the students and professors work together and learn from each other." They stated that a very important enabling factor was also that "the professors did not have any background knowledge on building the low-cost EEG headsets or eye trackers," thus reducing the knowledge gap by forcing the participants to work together to solve the design challenge. The workshop also impacted individual career paths; for example, Hermes Huang encouraged one of the workshop organizers to apply for a Mozilla Open Science Fellowship, which enabled him to continue post-doctoral research at Stanford University, where he is continuing to promote Open Science.

The last workshop in Nepal showed a similar dynamic between the transnational nature of OSH and localized tinkering. The workshop, which was seed-funded through the OCSDNet project, won further support from the Danish Center of Culture and Development (CKU) and the Center for Social Development Studies at Chulalongkorn University. The ten-day workshop's main deliverable was to co-develop a space technology curriculum for young students. Over the ten days, three teams co-created and tested hardware-based kits for DIY antennas, water rockets, and "life in space."

OSH Transnational Challenges and Opportunities in Southeast Asia

In all three countries that were part of our project, we saw a similar pattern, where a visitor from abroad (Marc Dusseiller in the case of HONF and Lifepatch, but also numerous others in Indonesia) or a "transnational" geek returning to his home country after studying abroad (in Nepal and Thailand) and foreign expats living there (Thailand) became involved with the local organizations or created their own spaces. The crucial role of these cosmopolitan individuals was very visible in the first Thai workshop on open hardware IoTs for making and hacking. The majority of the ten participants were either US-educated Thais or regional and international expats of Nepalese, US, or Canadian origins who settled in Thailand. Two of them were already prolific makers, with their 3D printing and electronics projects involved in agricultural innovation and health having been featured by international media.

The example that summarizes the transnational phenomena of OSH in Thailand is Raitong Organics Farm,[11] a social enterprise founded by a Thai and South African couple. In addition to running their own business and farm, they support other farmers around the country by engaging with hacker and maker communities and design thinking and social innovation organizations to create low-cost technology solutions and training programs. The innovative farm experiments with sensor technologies for improving crop quality regularly cooperates with students from around the world on various DIY ideas (Figure 3.5). They are part of an international movement of similar "hacker farms," like the one near Tokyo started by Akiba, who used to run the Tokyo hackerspace before relocating to the countryside. Raitong Organics Farm has since become an innovation centre supported by the International Development Innovation Network housed at D-Lab at the Massachusetts Institute of Technology.

Figure 3.5. International Development Design Summit at Raitong Organics Farm.

Photo credit: Deborah Tien.

The OSH projects, which we initiated and followed, raise an important challenge regarding how to utilize or formalize these transnational and cosmopolitan exchanges already happening on the ground. Can we claim that these Open Science activities create a more global and cosmopolitan science outside the national (and nationalistic) policies measuring innovation and research purely on the number of local patents and citations? Should these transnational dimensions of knowledge exchanges over open source technologies and models of work be supported by the national policies? They certainly informed our category of "openness" in the workshops as global, international, and even transnational exchange and cooperation that aspires for more equal ground rather than only to "diffuse" a technology. These transnational aspects of OSH projects were partially in tension with the Indonesian projects, which emphasize more the tacit knowledge and "little science" collaborations.

The exchanges we witnessed on the ground were mostly multidirectional rather than symmetrical or unidirectional: for example, a project that was initiated in Indonesia by a foreign visitor from Switzerland or the EU in collaboration with participants from Indonesia, India, and Nepal, and which was then continued and developed further in India (the example mentioned is from HackteriaLab 2014, Yogyakarta workshop[12]). The transnational networks surrounding OSH activities are rather idiosyncratic without clear geopolitical and economic logic, functioning more as networks of friendships. These cosmopolitan aspects of the OSH activities and development were more clearly embraced in Thailand, while in Indonesia and Nepal we witnessed some attempts to emphasize also the local origins of tinkering, making, crafts, and knowledge production. This was surprising because Thailand had the most developed open hardware ecosystem in terms of manufacturing, so its local capacity to produce OSH was strongest, but not the most developed.

The importance and emphasis on the local OSH activities and tinkering related closely to the type of content produced in the workshops. The projects in Indonesia supported more artistic and design-oriented OSH practices and collaborations, which were more embedded in the local communities and involved crafts. In Thailand, the OSH workshops supported a more globalized and cosmopolitan notion of knowledge sharing, emphasizing the educational and research functions of such tools (Little Bits workshop or the open EEG-electroencephalogram tools). Similarly, in Nepal, the workshops supported educational and entrepreneurial goals. The tensions between the transnational aspect of OSH as infrastructure and OSH as a tool to enhance existing tinkering practices was also visible on the level of the language and concepts used by the participants to describe their practices. The concepts of making, hacking, and do-it-yourself (DIY) alongside related terms such as design thinking and innovation played important roles in all three sites. The organizations in all three countries embraced, to various degrees, combinations of these concepts as part of their mission statements.

While in Indonesia the work on OSH was an opportunity to celebrate and embrace the local practices of tinkering; in Thailand and Nepal, there was a stronger need to identify with the more global and universal terms of the maker and hacker movement. For example, the FabCafe, one of the sites of our workshops in Bangkok, Thailand, combines the ideas from the Massachusetts Institute of Technology (MIT)-based project of Fab Labs, where "you can make your idea into the reality with digital fabrication tools," with the idea of a network of cafes around the world that enable cooperation between makers. The FabCafe self-identifies with the FabLab global community, which claims that it is "not only a 'digital fabrication cafe,' but also a 'local design community' and 'global business network'." We believe that our community will bring innovation into the future of making!" (www.fabcafe.com). In Nepal, the Karkhana collective is described as something between a makerspace and San Francisco's Exploratorium: "an education company and makerspace with a unique approach to learning" where "Our teachers...turn the classroom into a lab for discovery" (www.karkhana.asia).

Only in the case of Yogyakarta, Indonesia, did we notice an emerging resistance to these terms in 2015, which was first described by a researcher, Cindy Lin (2015), who wrote her honours thesis on the local expressions of making and tinkering and organized an exhibition in Yogyakarta on the topic in March 2015. The reason for this resistance to the American ideas of making and hacking is the rich local vocabulary of expressions that describe making and tinkering in Indonesia, as well as the local history of the citizen science organizations that are very much involved with various community projects either in the universities or their neighbourhoods. In the case of our main partner, HONF, which started its activities in the late 1990s, the reason is also the colonial heritage, which preserved stronger links to Europe rather than the US (Kera 2012). In the case of the Lifepatch citizen science group operating in the Bugisan neighbourhood of Yogyakarta, the opposition to the terms hacker and maker was explicit even in the descriptions of their projects. They used local concepts of tinkering, crafts, and making, but also "engineering" (as discussed by Cindy Lin) to refer to the local secondary vocational education (Lin 2015).

The tension between the international terms of making and hacking and the notions and practices of tinkering, crafts, and community organization is the reason why we decided to define "openness" in our project as exchange on the level of tacit knowledge rather than only transnational Open Science models of sharing. It is a type of OSH activity that preserves the connection to the local origins of tinkering and everyday practices. Making tools as a way of discussing what is science and how it related to local ways of living and doing things proved to be the most valuable experience for our participants in the first workshop and inspired the design of the rest.

Conclusion

The examples of workshops and OSH instruments held during our project explore a notion of Open Science as "little science" that is situated in specific contexts and community, supporting local interests and practices related to food, agriculture, fashion, education, etc., without a clear hierarchy. OSH as a model for the Global South simply enables people to experiment with various aspects of their everyday activities and rethink the meaning and influence of science upon their community without accepting the goals of "big" and professional science. We call these unique situational and everyday practices and engagements surrounding OSH projects "tinkering" and "little science." While these community-based engagements around OSH support learning within a particular context and community, they also intersect with larger issues of educational infrastructure, access to technology, and equal valuation of people's time, knowledge, and experience, regardless of their background.

"Little science" has the opportunity to create a body of foundational knowledge that can begin to address these larger issues while creating impact in local communities. However, it will take profound leadership and vision from diverse stakeholders to utilize the variety of data created and expressed in "little science" workshops, engagements, and methods. In this, we see an interesting tension between the more transnational goal of OSH instruments and Open Science trying to create standards and infrastructure that bridge technological and other divides and our experience with the workshops that support "little science" engagements over OSH with diverse and hybrid agendas. The hybrid and diverse OSH agendas at our three sites combine the transnational goals with local needs and ideas, but they also have the freedom to question and even refuse the connection of OSH to "big science" and industry needs.

Notes

1. See http://openhardware.science/about/why-gosh/ for more information on GOSH.
2. See http://www.appropedia.org/Category:Open_source_scientific_hardware for more information on OSAT.
3. See http://www.gaudi.ch/GaudiLabs/?page_id=328 for more on Hackteria's "Generic Lab Equipment."
4. See https://www.thingiverse.com/jpearce/collections/open-source-scientific-tools /page:1 for more on the Thingiverse collection.
5. See http://www.karkhana.asia/stories/k_space-team-rocket/ as an example.
6. Find and learn more about Raspberry Pi, a small, more affordable computer that can be used to learn programming here: https://www.raspberrypi.org/.
7. Find and learn more about Arduino, an open-source electronic prototyping platform here: https://www.arduino.cc/.
8. Find Vicky's consulting profile at UpWork here: https://www.upwork.com/o /profiles/users/_~012ed5a0325b0df5c8/.
9. Learn more about Cobanana here: http://www.cobanana.com.
10. See https://twitter.com/htkhuang/status/701623029894356992 for the full original call online.
11. Learn more about Raitong Organic Farms here: https://www.facebook.com /RaitongOrganicsFarm/.
12. Learn more about Hackteria Lab 2014 here: http://wlu18www30.webland.ch/wiki /HackteriaLab_2014_-_Yogyakarta.

References

Bessen, James, and Michael James Meurer. 2008. *Patent Failure: How Judges, Bureaucrats, and Lawyers Put Innovators at Risk*. Princeton, NJ: Princeton University Press.

Birtchnell, Thomas, and William (William Anthony) Hoyle. 2014. *3D Printing for Development in the Global South: The 3D4D Challenge*. Basingstoke, UK: Palgrave Macmillan.

Blikstein, P. 2013. "Digital Fabrication and 'Making': The Democratization of Invention | Transformative Learning Technologies Lab." In *FabLabs: Of Machines, Makers and Inventors*, edited by J. Walter-Herrmann and C. Büching, 203–22. Bielefeld, Germany: Transcript Publishers. https://tltl.stanford.edu/publications/papers-or-book-chapters /digital-fabrication-and-making-democratization-invention.

Bock, Gregory, and Jamie Goode. 2007. *Tinkering: The Microevolution of Development*. Chichester, UK: John Wiley & Sons. https://doi .org/10.1002/9780470319390.

Borgman, Christine L., Jillian C. Wallis, and Noel Enyedy. 2007. "Little Science Confronts the Data Deluge: Habitat Ecology, Embedded Sensor Networks, and Digital Libraries." *International Journal on Digital Libraries* 7 (1–2): 17–30. https://doi.org/10.1007/s00799-007-0022-9.

Busch, Peter, and Debbie Richards. 2004. "Tacit Knowledge and Culture." In *People, Knowledge and Technology: What Have We Learnt so Far? Proceedings of the First iKMS International Conference on Knowledge Management*, 187–98. Singapore: World Scientific Publishing Co. Pte. Ltd. https://doi .org/10.1142/9789812702081_0018.

Carillo, Maria Rosaria, and Erasmo Papagni. 2014. "'Little Science' and 'Big Science': The Institution of 'Open Science' as a Cause of Scientific and Economic Inequalities among Countries." *Economic Modelling* 43: 42–56. https://doi.org/10.1016/j.econmod.2014.06.021.

De', Rahul. 2015. "Open Source Software in the Global South." In *The International Encyclopedia of Digital Communication and Society*, 1–9. Hoboken, NJ: John Wiley and Sons, Inc. https://doi.org/10.1002/9781118767771 .wbiedcs024.

Dickson, David. 1988. *The New Politics of Science*. Chicago: University of Chicago Press.

Egghe, L. 1994. "Little Science, Big Science...and Beyond." *Scientometrics* 30 (2–3): 389–92. https://doi.org/10.1007/BF02018109.

Fienberg, Richard Tresch, Pamela L. Gay, Gary Lewis, and Michael Gold. 2011. "Citizen Science Across the Disciplines." *Earth and Space Science: Making Connections in Education and Public Outreach* 443: 13–23.

Forero-Pineda, Clemente. 2006. "The Impact of Stronger Intellectual Property Rights on Science and Technology in Developing Countries." *Research Policy* 35 (6): 808–24.

Franzoni, Chiara, and Henry Sauermann. 2014. "Crowd Science: The Organization of Scientific Research in Open Collaborative Projects." *Research Policy* 43 (1): 1–20. https://doi.org/10.1016/j.respol.2013.07.005.

Gascoigne, Neil, and Tim Thornton. 2013. "Tacit Knowledge." *Tacit Knowledge*. Durham, NC: Acumen Publishing.

"Global Open Science Hardware (GOSH) Manifesto." 2016. *Gathering for Open Science Hardware*. http://openhardware.science/gosh-manifesto/.

Gortych, Joseph E. 2014. *Consider a Spherical Patent: IP and Patenting in Technology Business*. Boca Raton, FL: CRC Press.

Griffiths, Sarah. 2013. "The Amateur at Play: Fab Labs and Sociable Expertise." In *Crafting the Future: 10th European Academy of Design Conference*, 1–7.

Gura, Trisha. 2013. "Citizen Science: Amateur Experts." *Nature* 496 (7444): 259–61.

Haunss, Sebastian. 2013. *Conflicts in the Knowledge Society: The Contentious Politics of Intellectual Property*. Cambridge: Cambridge University Press.

Huang, Hermes. 2015. "Networks of Practice Around Open Science: A Case Study on The House of Natural Fiber Foundation in Yogyakarta, Indonesia." Master's thesis, Chulalongkorn University, Bangkok, Thailand.

Kaiying, Cindy Lin, and Silvia Lindtner. 2016. "Legitimacy, Boundary Objects and Amp; Participation in Transnational DIY Biology." In *Proceedings of the 14th Participatory Design Conference on Full Papers - PDC '16*, 171–80. New York: ACM Press. https://doi.org/10.1145/2940299.2940307.

Kera, Denisa. 2012. "Hackerspaces and DIYbio in Asia: Connecting Science and Community with Open Data, Kits and Protocols." *Journal of Peer Production* 2: 1–8. http://peerproduction.net/issues/issue-2/peer-reviewed-papers/diybio-in-asia/?format=pdf.

——. 2014. "Innovation Regimes Based on Collaborative and Global Tinkering: Synthetic Biology and Nanotechnology in the Hackerspaces." *Technology in Society* 37 (1): 28–37. https://doi.org/10.1016/j.techsoc.2013.07.004.

——. 2015. "Open Source Hardware (OSHW) for Open Science in the Global South: Geek Diplomacy?" In *Open Science, Open Issues*, edited by Sarita Albagli, Luca Maciel, and Hannud Akexandre Abdo, 133–57. Brasilia: Instituto Brasileiro de Informação em Ciência e Tecnologia (IBICT). http://livroaberto.ibict.br/handle/1/1061.

Lee, Joo-Young. 2015. *A Human Rights Framework for Intellectual Property, Innovation and Access to Medicines*. Abingdon: Routledge.

Lewenstein, Bruce V. 2004. "What Does Citizen Science Accomplish?" Meeting on Citizen Science, Paris, June 8, 2004.

Lievrouw, Leah. 2010. "Social Media and the Production of Knowledge: A Return to Little Science?" *Social Epistemology* 24 (3): 219–37. https://doi.org/10.1080/02691728.2010.499177.

Lin, Cindy. 2015. "Indonesian Hacking: Local Particularities and Global Interactions of Hackerspace in Indonesia." National University of Singapore.

Lindtner, Silvia, Anna Greenspan, and David Li. 2015. "Designed in Shenzhen: Shanzhai Manufacturing and Maker Entrepreneurs." *Aarhus Series on Human Centered Computing* 1 (1): 12. Aarhus, Denmark: Aarhus University Press. https://doi.org/10.7146/aahcc.v1i1.21265.

Livingston, Dennis. 1976. "Little Science Policy: The Study of Appropriate Technology and Decentralization." *Policy Studies Journal* 5 (2): 185–92. https://doi.org/10.1111/1541-0072.ep11795258.

Maclurcan, Donald, and Natalia Radywyl. 2012. *Nanotechnology and Global Sustainability*. Boca Raton, FL: CRC Press.

Mellis, David A., and Leah Buechley. 2011. "Scaffolding Creativity with Open-Source Hardware." In *Proceedings of the 8th ACM Conference on Creativity and Cognition - C&C '11*, 373–4. New York: ACM Press. https://doi.org/10.1145/2069618.2069702.

Moilanen, Jarkko. 2012. "Emerging Hackerspaces - Peer-Production Generation." In *IFIP Advances in Information and Communication Technology* 378: 94–111.

Moore, Kelly, Daniel Lee Kleinman, David Hess, and Scott Frickel. 2011. "Science and Neoliberal Globalization: A Political Sociological Approach." *Theory and Society* 40 (5): 505–32.

Neylon, Cameron, and Shirley Wu. 2009. "Open Science: Tools, Approaches, and Implications." *Pacific Symposium on Biocomputing* 14: 540–4.

Nowotny, Helga, Peter Scott, and Michael Gibbons. 2001. *Re-Thinking Science: Knowledge and the Public in an Age of Uncertainty.* Cambridge, UK: Polity Press. http://www.polity.co.uk/book.asp?ref=9780745626079.

Nutch, F. 1996. "Gadgets, Gizmos, and Instruments: Science for the Tinkering." *Science, Technology and Human Values* 21 (2): 214–28. https://doi.org/10.1177/016224399602100205.

Open Source Hardware Association. 2017. "Definition (English) – Open Source Hardware Association." https://www.oshwa.org/definition.

Pearce, Joshua. 2012. "Building Research Equipment with Free, Open-Source Hardware." *Science* 337 (6100): 1303–4.

———. 2014. *Open-Source Lab: How to Build Your Own Hardware and Reduce Research Costs.* Waltham, MA: Elsevier. https://doi.org/10.1016/B978-0-12-410462-4.01001-5.

———. 2015. "Quantifying the Value of Open Source Hard-Ware Development." *Modern Economy* 6 (1): 1–11. https://doi.org/10.4236/me.2015.61001.

Price, Derek John de Solla. 1986. *Little Science, Big Science—and Beyond.* New York: Columbia University Press.

Sillitoe, Paul. 2007. "Local Science vs. Global Science: An Overview." In *Local Science vs Global Science: Approaches to Indigenous Knowledge in International Development,* edited by Paul Sillitoe, 1–22. Oxford, NY: Berghahn Books.

Smith, A., C. Lintott, and Citizen Science Alliance. 2010. "Web-Scale Citizen Science: From Galaxy Zoo to the Zooniverse." In *Proceedings of the Royal Society Discussion Meeting "Web Science: A New Frontier."* London: The Royal Society.

Sobkowicz, Pawel. 2011. "The Role of Internet in Citizen Science." In *Internet Policies and Issues,* Volume 9, edited by B. G. Kutais. New York: Nova Science Publishers.

Takhteyev, Yuri. 2012. *Coding Places: Software Practice in a South American City.* Cambridge, MA: MIT Press.

CHAPTER 4

On Openness and Motivation: Insights from a Pilot Project in Latin America

Josique Lorenzo, John Mario Rodriguez,
and Viviana Benavides

Abstract

This chapter reflects on the importance of understanding the motivations of participants and non-participants for engaging (or not) in small-scale citizen science projects that require a high level of voluntary engagement. The multi-motivational nature of involvement in such initiatives is illustrated through the experience of a pilot project implemented in rural communities of Costa Rica and Colombia, thereby contributing to our understanding of the challenges and opportunities associated with designing locally and culturally appropriate research initiatives.

Introduction

This chapter examines the experience of a pilot project conducted in Costa Rica and Colombia from 2015–2017. The project was built on participatory approaches with the goal of fostering collaboration between academic representatives and local communities. Success in that regard has been uneven. Strong motivation based on multiple types of goals or motives was essential for effective and ongoing participation by the teams and individuals. Hence, the purpose of this chapter is to make sense of and reflect on the motivations for engaging, or not engaging, in what was a small-scale, high-involvement project, thereby setting foundations for future reflection and research.

In this chapter, we first briefly introduce our use of the concepts of openness and motivation. We then provide an overview of the project and comment on the role of motivation in that context. We close the chapter by presenting key insights gained from this experience.

On Openness and Motivation

There are many ways to define openness in context; reflecting on the project's experience will contribute to defining one of many possible types of openness. In the particular context of the project presented in this chapter, openness is viewed as a mindset, a state of mind, or attitude that is adopted primarily by individuals. Openness calls for a commitment to difference, to rationalizing and doing things differently, and to adopting a self-reflective, critical practice. It is rooted in a broad conception of science that allows for the expression of such difference and valorizes other ways of learning and knowing.

Openness means, among other things, to actively communicate the scientific knowledge to non-traditional audiences: for instance, rural dwellers often assist researchers and students in performing their work (by providing information, participating in surveys, and so on), but researchers and students do not typically feel bound to share their results with rural communities, or tend to do so without taking proper care to translate their message.

Open Science is sometimes referred to as "community science." In the context of this project, research began and ended with community problems, rather than with scientific problems. Thus, during the first work session, the participants worked on their own collective definition of the topic brought by researchers (in this case, climate change adaptation) and identified what concrete problems there were in their community or area in relation with it. This served as a basis for subsequent development of concrete knowledge to be investigated, skills to be strengthened, and ideas regarding solutions that could be implemented.

Community participation in natural resources management is critical because the objectives of conservation or sustainable management do not always coincide with community or social objectives. In this project, we used the term "community," as opposed to academia, to refer either to spatial units characterized by their smallness and territorial attachment (e.g., a village) or to groups that share a certain set of norms or practices (e.g., an association of citizens).[1]

The Role of Motivation

Participatory methods that emphasize participation and action by local communities, such as participatory action research or participatory mapping, are at the core of open practices such as citizen science, and they have a long history behind them. These approaches have been used and abused, at times glorified for their potential for social transformation and at other times discredited or accused of bringing about a new form of tyranny (Cooke and Kothari 2001). According to Robert Chambers (2006), "we have now entered a phase of increasingly inventive and eclectic pluralism with borrowing and cross-fertilization between participatory streams." This new phase is both extremely enriching and complex.

The reasons behind a decision to participate or not in projects that require a high level of voluntary engagement are also multifaceted and diverse. Understanding motivation—defined as the general desire to do something, and, in plural, as the goals that energize and direct behaviour—could allow us to improve the design of participatory projects in the future and ensure their sustainability over time. As noted by Rotman et al. (2012), two pivotal points in participation are significantly affected by motivational factors: (1) the initial decision to participate in a project; and (2) the ensuing decision to continue. These factors could be especially critical in projects in which community groups are placed at the centre of sustainability and play an active role in all phases.

Our project builds on existing work looking at motivation and its characteristics as intrinsic or extrinsic, i.e., driven by internal rewards such as a desire for personal improvement, to learn, or to feel accomplished, or by external rewards or constraints such as a desire to impress or to receive a monetary award (Tyler 2010; Kirkland et al. 2011; Ryan and Deci 2000; Deci and Ryan 2000; Batson et al. 2002). In this chapter, we build on this work to nurture our own understanding of the motivational process.

The Project

The goal of the two-year pilot project, "Improving Adaptive Capacity Through Open Collaborative Science: A Case Study in Two Model Forests," was to improve both the human capabilities and knowledge capital of individuals and local communities while at the same time stimulating the creation of new social ties. The project was a small-scale/high involvement initiative that started in March 2015 and ended in February 2017. During this period, more than fifteen focus groups, workshops, and field trips were conducted. These were complemented with phone calls and follow-up visits. In total, over thirty participants were directly involved in the project at one point or another.

The project was conducted in Colombia and Costa Rica, two biodiversity-rich countries. In 2012, Costa Rica and Colombia were rated as the first and third "happiest" countries of the world, respectively, according to the Happy Planet Index, an alternative measure of sustainability. Specifically, the project was conducted in the territories of two Model Forests: Reventazón in Costa Rica and Risaralda in Colombia.

Model Forests are social platforms in which people participate voluntarily, working in partnership toward a common vision for the sustainable development of a large landscape with rich and abundant natural resources, including but not limited to forests. The Canadian government initially introduced the concept at the Earth Summit in Rio in 1992 as a way to promote multi-stakeholder conflict resolution.[2] The platforms have evolved since then to include many other types of activities related to natural resources management and conservation. Just like UNESCO's biosphere reserves, Model Forests are areas of "recognition" rather than "regulation," which means that their presence does not alter formally the configuration of laws, policies, or property rights over landscapes. However, they have

> provided opportunities for engagement of local people in environmental issues, networking with other actors on common agendas, providing demonstration areas for specific kinds of research or development priorities, and serving an honest broker function to advance specific initiatives. (Gerardo et al. 2017)

The Model Forest concept is based on sharing knowledge and on a sense of community. It seeks to promote a constructive and open dialogue between competing land uses as well as a culture of collaboration, engagement, and participation, generally with the leadership of grassroots organizations. These platforms seemed like an ideal starting point to implement a pilot project that promoted Open Science and citizen engagement.

The project focused on two types of actors: on one hand, the scientists or academic researchers, and, on the other hand, the local representatives or leaders of different communities. Both groups of actors were characterized by heterogeneity. There were an almost equal number of men and women. Scientists were represented by master's students and professors/researchers who had varying degrees of experience in engaging directly with communities and who expressed interest in participating. Community teams comprised three to six people, mostly adults, but some of them also included a few teenagers and older people. All of the teams had at least one leader or person who was very engaged in the local community with experience participating in social platforms.

The community groups were intentionally selected to represent different parts of the Model Forest landscape, with diverse biophysical and socioeconomic conditions.[3] There is a crucial relationship between biodiversity and productive systems in rural landscapes: communities face the challenge of producing more without destroying the natural capital and ecosystem services, at the same time taking into account the threats posed by climate change. So, in Colombia, the project involved, among others, citizens from three different areas along the same watershed (upper, middle, and lower basin of the Otún River), which allowed for recognition of the interconnectedness of the issues they all might face. For example, contamination upstream can have consequences for water users downstream. (It should be noted that the Otún River is the only source of drinking water for approximately half a million people.) In Costa Rica, the communities selected to participate were located on an altitudinal gradient (elevations within the Reventazón Model Forest range from 410 to 3,500 metres above sea level), which provided an opportunity for them to understand the diversity of ecosystems and climate change impacts in different life zones (such as premontane and montane rainforests).

The core topic addressed was climate change adaptation. It should be noted that there is a growing body of knowledge in the field of community-based adaptation (CBA), which builds on values and approaches similar to the ones promoted within the project. However, even though locally initiated and led projects are an important dimension in community-based adaptation, they should not always be assumed to be the best.[4] In our case, the framing of climate change adaptation did not feel contrived to participants since the topic was sufficiently broad and our approach flexible enough to include interconnected and relevant topics of direct interest to local teams.

Methods

The approach used was based on classic Participatory Action Research (PAR),[5] but it sought to integrate a more empowering or "extreme" citizen science component inspired by methodologies such as the one developed by the International Center for Tropical Agriculture [CIAT] in the 1980s (see Ashby et al. 2001). Moreover, it included the notion of "reciprocity" (Brereton et al. 2014), that is, the importance of not only taking but also giving back to the community. In other words, the researchers had the intention to work with community stakeholders: the aim was to widen their horizons and give them opportunities for self-organization, including defining the problems and establishing local priorities. This was accomplished through a series of work sessions, meetings, and field trips during which we used a mix of tools and materials that we either developed or adapted (e.g., drawing maps, games, and classic PowerPoint presentations on scientific topics as shown in Figure 4.1 facing page).

Community participants, in collaboration with academic representatives and other partners, proposed and subsequently implemented seven ideas of locally relevant adaptation initiatives. Table 4.1 (page 94) gives an overview of the seven micro-initiatives that have been designed as part of the project. These were presented during final events in both Model Forests where members of different institutions were invited.

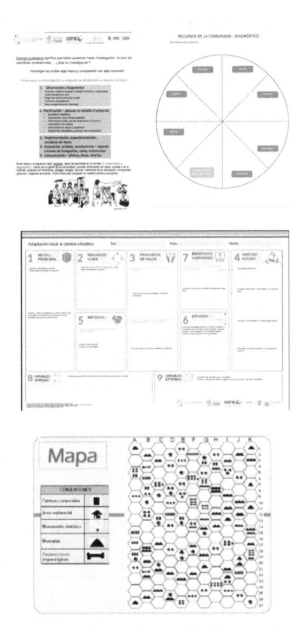

Figure 4.1. Examples of printed materials that were developed/ adapted to conduct the work sessions: From left to right, a guide with steps for participants to engage in the project, a community capitals analysis sheet, a lean canvas, and a game on common goods called Mapa.

Table 4.1. Seven local initiatives created as part of the project

Name of local initiative	Team/community	Summary/goal of the initiative
1. Creation of an agroecological network	Young men and women from Belen de Umbria, and surroundings, Colombia, who have decided they needed to innovate to change their communities. They are strongly engaged in the local coffee committee.	The team created their own foundation to manage the network. They were successful in securing the support of several institutions that helped them to set up a website to disseminate agroclimatic information among producers. They also investigated the use of other ICTs to this end. (For example, they are considering sending text messages.) Their network has already attracted several new members. They also initiated a beekeeping project that will make their network self-sustainable.
2. Ecotourism and preserving the historical memory in Villa Mills	ASOPROFOR (Association of Forest Producers), Villamills, Costa Rica. The leader of the group possesses an extensive knowledge of the wildlife of the surrounding primary forest and has collaborated with the academic institution for years.	The team decided that it was important to "retrieve" the historical and collective memory of this small and remote town often forgotten (the highest and coldest in Costa Rica)—for example, by collecting old photographs and testimonies from older people in the village and digitizing these. They actively engaged in training with digital technologies and capacity building to improve their presentation skills. Finally, they put emphasis on promoting ecotourism to educate people about biodiversity in those high altitudes.

Name of local initiative	Team/community	Summary/goal of the initiative
3. Green Paths (*Caminos verdes*)	Residents of Cerritos. One resident is the private owner of the only protected area of tropical dry forest in the Risaralda Model Forest, and another one operates one of the first plantations of Guadua agustifolia (bamboo) in the world that has been certified by the Forests Stewardship Council (FSC).	The goal of the initiative is the establishment of a biodiversity restoration plan to improve connectivity between forest patches located in a semi-urban area. The initiative is being conducted in an area that has been impacted negatively by real estate development. The team members have conducted a diagnostic through satellite maps and engaged with filmmakers to prepare a drone video of the area for dissemination. They have also established a tree nursery and started educational and tree planting activities with the schools in the area.
4. Nursery of Volcanic Life (*Vivero de Vida Volcánica Turrialbeño*)	Members of the Northern Biological Subcorridor (*Subcorredor Norte*), living in Santa Cruz de Turrilaba, Costa Rica. The team had participated actively in the first designation of origin to be obtained by a Central American dairy product (Turrialba cheese).	The goal of the initiative was the creation of a tree nursery to reforest the area (located near a volcano) and help attract birds in danger of extinction, such as the quetzal. The initiative involved their knowledge management and further research about tree species suited to high altitudes, wet climates, and volcanic soils. They also plan a partnership with a public institution that is interested in reforesting the area and buying the trees. There are many challenges to overcome; for example, since the start of the project, the volcano has become increasingly active, and the ashes have destroyed seedlings.

Name of local initiative	Team/community	Summary/goal of the initiative
5. Rainwater harvesting	Agroecological School of Santa María de la Loma. Small group of women, who are agroecological producers and who are interested in water as a fundamental element of life. This initiative is part of their daily life.	Creation of a rainwater harvesting strategy for a community of producers with two demonstration sites, in order to support organic production for both self-sufficiency and income generation. This involved documentation and research regarding the best system to use, given the conditions of the housing in that area. It also had an educational component as they promoted "water culture" through the activities organized by their local agroecological school, a farmer-to-farmer initiative that organizes meetings several times a month. The main climate change impact addressed was the more frequent droughts in their area.
6. Cultural and educational strategy for responsible consumerism	Leaders of three communities (*veredas*): La Bananera, La Florida, and those who have been very actively engaging communities in waste management through innovative local programs and who have created a cineclub to stimulate cultural life in the area.	Taking into account climate threats, the aim of this initiative is to educate the local populations of the watershed and visitors, especially those coming from the urban area who use the areas along the river during the weekends for recreational purposes. The initiative is focused on sustainable consumption, waste management, and avoiding contamination of the river. It involved an extensive survey and workshops among the communities to decide which specific aspects to focus on, including what animal or plant would best represent the area as a "mascot" for the educational campaign.

Name of local initiative	Team/community	Summary/goal of the initiative
7. TurriAbonos	Members of APOYA, Costa Rica, a local association of organic and agro-sustainable producers. The team members had little to no experience in tackling such an initiative before, but the Association has been supporting them.	Production of organic fertilizers to prevent contamination of the soils and enhance production. Through different tests and chemical analyses in a laboratory, the team is seeking the best formula to improve productivity of home gardens while keeping the soils healthy. The goal was to start an agri-business to make the initiative self-sustainable and generate income by selling the fertilizers first to local organic producers and members of the association in order to eventually get the product certified. The project involved market research and acquiring specific entrepreneurial skills.

One study suggests that activities devised by participants themselves may have a better chance to lead to long-lasting commitment (Dickinson et al. 2012). Time will tell if these local initiatives will keep moving forward on their own. In many cases, community members made concrete plans for making their initiative sustainable (for example, selling products to recover ongoing future costs or seeking new partnerships); in other cases, securing more funding will be crucial. Even though some of these initiatives may not last, the learning achieved during the process and the skills gained cannot be unlearned or undone. That brings us to an important lesson: we tend to focus much more on outcomes than on processes, but, in our conception of Open Science, the road taken is as important as the destination.

The Role of Motivation in the Project

At the beginning of the project, teams were selected based on their motivation, which was assessed by conducting a series of personal visits to individuals representing local communities that had been previously identified either through their participation within the Model Forest platform or through contacts of the Model Forest participants. The teams were required to commit to the entire process, which included actively engaging in work sessions and in the elaboration of an initiative related to climate change adaptation. Incentives included opportunities to travel to new areas of their territory, make new connections, design their own initiative with the help of the researchers, and receive a modest seed fund to give it a kick-start.

Motivations were identified through a mix of observation, conversations, and direct feedback. The participants were asked basic questions during informal activities such as follow-up visits, including their reasons for participating, whether they were still motivated and why, what they thought could be achieved through their participation, and so forth. Workshop evaluations also provided an opportunity to better understand what motivated people.

Three out of ten community teams that were initially selected dropped out early in the process, emphasizing the importance of understanding motivation (and "amotivation").[6] The remainder displayed a high level of motivation during the whole process.[7] General observations include the following:

1. Despite the external reward provided, intrinsic motivation was key, including learning, meeting new people, and simple enjoyment. Indeed, all participants displayed a high degree of self-determined motivation, associated with the three psychological needs previously mentioned (autonomy, competence, relatedness). They kept asking when the next meeting would take place.
2. The level of motivation seemed to decrease whenever there was less frequent follow-up from the academic researchers. The feedback given by peers was also an important motivational factor.
3. Motivation was primarily related to the opportunity to tackle challenges from a non-traditional perspective, through an innovative and flexible approach. People reported feeling responsible for the project, having freedom to act, and being satisfied with the fact that their opinion mattered.
4. The leadership of one or two individuals was essential to motivate the rest of the group. The members of the teams

were people who already knew each other for a long time and trusted each other, which was an important factor for having them work effectively together. There were two cases where internal disagreements among team members de-motivated some participants.

The project was more successful in enabling meaningful dynamics that were appropriated by the local teams than in generating enthusiasm and engaging academic researchers. Indeed, several researchers did not sustain their interest in the initiative over time. Reasons for this could include that the perceived benefits of participation were greater for the local teams than for the scientists; the lack of academic incentives can be un-motivating; or that, generally speaking, researchers have access to a wide range of research projects (including paid opportunities) that seem more interesting to them for a range of reasons. However, for many local groups, this project was seen as quite a unique opportunity to act. Broad generalizations include:

1. Early career professionals and students tend to be more open to the idea of openness and of investing significant time in such a project. (However, academic institutions generally value experience over good ideas.)
2. Several professors considered it important to participate in a project that was built on citizens' perspectives over the landscape as a means to achieve a stronger articulation between the objectives of the people and the objectives of conservation.

Below, we enumerate some of the motivations of both the scientists and the citizens. In reality, however, these were not so clear-cut and evolved throughout the research cycle. Moreover, the boundaries between groups are sometimes blurred or fluid, that is, we tend to forget that scientists are also citizens themselves. In the case of the project, there were even participants who were not exactly part of either of those two teams—for example, a foundation that works on design and innovation that was not part of the academic institution nor of the community but played a facilitating role in enabling the process. Another example is the case of two students in Colombia who were actually part of a team in their role as citizens.

Academic researchers were motivated, for example, by the fact that they could enhance and validate their own research or that of their group (egoistic and collectivist motivation), or that they could help others to better understand certain challenges and give back to the community (altruistic motivation). They generally held the principle that good science ought to be useful to society, and they trusted the locals as being the "experts." On the other hand, some scientists who refused to engage on a voluntary basis decided that the process was too time-consuming or not worth the effort. According to them, the contributions that community members could make were rather limited and such a process was unnecessary, in their perception, to achieve the research goals in an efficient manner.

Local citizens who participated were prompted by the desire to learn and improve themselves as well as their community, to meet new people, and to self-organize. They had a keen desire to better understand the research conducted in their area and hoped there would be follow-up afterwards. They were confident that the process could help them engage in new practices and tended to be proactive and curious. Those who chose not to participate were, in some cases, simply not interested in the topic, felt "consultation fatigue," or considered the workshops a waste of time.

Of course, other factors, such as self-esteem or practical issues, are often at play in non-participation. For example, a highly motivated person may still not be able to engage due to personal circumstances (illness in the family, lack of time, etc.). Other factors could include culture or low educational attainment.[8]

Reflections on Lessons Learned and Recommendations

This experience illustrates the importance of being aware of the diverse array of motivations, and of making an effort to understand these in the context of a particular project. Most specifically in such high-involvement citizen science projects, taking into account what motivates people to participate can be key in developing mechanisms that ensure effective commitments and project success. In this regard, general reflections are provided below.

Promoting the right mindset: for more openness, there is a need to promote the right attitudes, which can happen even at an early age. An open mindset begins with an open education. Interestingly, within our project, four out of seven community-led initiatives had a specific educational component involving children. On the other hand, soft skills and values should be integrated in scientific training.

Values indeed always have an influence on the research (whether scientists recognize it or not), including our approach to the problem, our definition of the concepts, and the presentation of findings. In our project, the underlying values were those shared by the Model Forest network, such as respect of diversity or the pursuit of the common good.

Do not overvalue the label. Framing our approach with the label "Open Science" did not make a significant difference in working with the local teams, but it has helped in putting the approach forward for debate at the academic level, which is certainly important. Open Science advocates should remain open to the fact, however, that it is possible to do "Open Science" without calling it that. There are also many situations in which the "closed" way of doing science is still the best option possible; for example, in specific steps of the research cycle, when the analysis of data requires highly specialized skills, it might not be convenient to engage communities even if the analysis is conducted in their interest. Finally, we need to acknowledge that non-participation is also a valid choice that should be respected.

Feeling important is important. Using motivational tools is essential to sustain the level of engagement throughout a process like this one. The fact that people felt important and that their participation and opinions really mattered was crucial for continued engagement within the project. This included providing regular feedback as well as meaningful opportunities for meeting and engaging with peers and for taking initiative. Initial work to meet on a one-on-one basis was successful in engaging key individuals and clarifying expectations. It was also a good idea to provide a final opportunity to make a presentation in front of an external audience: one could feel how proud participants were of their accomplishments.

Time is golden. According to our experience, groups should have enough time for conscious reflection and to share the experience, within and outside of the work sessions. Sufficient time must be allowed between the sessions (several months). On the other hand, Robert Chambers (2006) pointed out quite rightly that taking people's time is often an abuse since time is precious for rural people, especially at critical times of the year (rainy season, harvest, etc.). Thus, it is important to adjust and establish a balance between the pace of the different stakeholders and efficiency goals (which are dear to researchers). In the case of our project, the opportunities given to local communities were a trade-off for the time investment they were willing to make. However, this was not so much the case for the academic researchers who also needed to invest a significant amount of their also precious time in the process. This aspect needs to be addressed by policymakers, donors, and academic institutions through appropriate incentives, such as recognition of the importance of this work or decent pay for researchers who are conducting it, including meaningful grants.

The main issues are not so much ones of scale, but of power. What really mattered in the work process was not how many people were engaged, but the fact that a more robust and horizontal relationship could be established between local people and researchers. Similarly, the external financial incentive offered to the community projects was low, but it still represented a meaningful opportunity in many senses, so it was more about having the power and autonomy to act and not so much about the scale of the initiatives per se. More extreme approaches to citizen science might be needed to effectively question power relationships between citizens and scientists (such as ExCites[9] or Comandulli et al. 2016, for example). Such approaches might be even more crucial when it comes to natural resource management, where decentralized approaches have proven to be key for introducing new social interactions between communities and their ecosystems, improving governance, and halting environmental degradation.

Do not underestimate the investment that is required. As shown in this pilot study, this type of project is more demanding than it seems. While many participatory projects aim at extracting information for the benefit of outsiders, the spirit here was to give back to the community and have a social impact. Bridging scientific work and what we could call "development work" requires an unusual amount of commitment. As learned through this experience, it is essential to plan in advance to devote the necessary resources, both human and financial, to ensure ongoing follow-up and feedback with local communities. Ford (2016) highlighted the "multifaceted role of the researcher" (as educator, communicator, facilitator, etc.) and the importance of negotiating relationships with community partners in a transparent manner from the onset. Indeed, this type of project calls for a specific profile or type of person or researcher—one who is deeply committed to the importance of integrating multiple forms of knowledge to understand today's problems, who is empathetic and communicative, and who is willing to interact with citizens on an equal footing—all of which involves sharing values, perspectives, and lifestyles, and doing so with deep respect. This is linked to the mindset and soft skills we mentioned above.

More support should be channelled into bottom-up approaches at all stages of the scientific process. The communities themselves should establish at least some of the priorities, which means leaving the agenda as "open" as possible. In the project, we let the teams start with defining the problems and then prioritizing, rather than coming with an analysis of the problem and a ready-made solution. Although this complicated the logistics, the fact that representatives of several communities were mixed together during workshops was enriching for all of them. Again, there is a need for more citizen science projects revolving around what is convenient for people and their lives and not only what is convenient for researchers. These types of projects should not be assumed to be better or worse; they are simply based on a different premise, including that the process of knowledge production is as important as the research outcomes—sometimes, even more so.

Conclusion

In September 2014, the terms "citizen science" and "citizen scientist" were added to the Oxford English Dictionary,[10] indicating the growing recognition of a phenomenon that is here to stay. It is interesting to note that the definition of "citizen scientist" accounts for both the role of the community and of the researcher in shaping their mutual relationships (note the comment in parentheses "now rare"):

> citizen scientist n. (a) a scientist whose work is characterized by a sense of responsibility to serve the best interests of the wider community (now rare); (b) a member of the general public who engages in scientific work, often in collaboration with or under the direction of professional scientists and scientific institutions; an amateur scientist.

As this new addition to the dictionary indicates, the boundaries between the roles of the citizen and the scientist are often blurred. In this chapter, we illustrated the multi-motivational nature of participation in community-based projects, providing initial reflections on a topic that deserves much attention since it shapes the collaborative relationships within citizen science projects relying on strong voluntary engagement. Our pilot project suggested that motivation that is internally generated is fundamental since external rewards, incentives, and financial support are generally scarce, low, or nonexistent.

Central to our approach was a focus on human capabilities and locally relevant development. Unlike many other citizen science projects, the project centred on community priorities, goals, and preferences. A lesson learned would be to more closely engage the scientific community throughout the process; in fact, many activities were devoted to making sure that the local citizen teams would participate, but comparatively fewer motivational efforts targeted those in academia.

Notes

1. According to Agrawal (1999), community is seen in three ways: as a spatial unit, as a social structure, and as a set of shared norms. It is on the basis of one or a combination of these three ideas that most of the advocacy for community rests. The concept of community as shared norms and common interests depends strongly upon the perceptions of its members; in this sense, all communities are "imagined communities."
2. Model Forests are members of the International Model Forest Network. For more information, visit http://www.imfn.net/.
3. The Model Forest is considered a "landscape approach." A landscape approach is "a conceptual framework whereby stakeholders in a landscape aim to reconcile competing social, economic, and environmental objectives" (*The Little Sustainable Landscapes Book*: http://globalcanopy.org/sites/default/files/documents/resources /GCP_LSLB_English.pdf).
4. For a critical reflection on CBA, see, for example, Ford et al. (2016).
5. The Participatory Action Research model begins with the interests of participants, who work collaboratively with professional researchers through all steps of the scientific process to find solutions to problems of community relevance. Finn (1994) outlined three key elements of participatory research: (1) it responds to the experiences and needs of the community; (2) it fosters collaboration between researchers and community in research activities; and (3) it promotes common knowledge and increases community awareness. http://www.ecologyandsociety. org/vol12/iss2/art11/.
6. Since this concrete project required a high level of initiative-taking and a rather long-term commitment, the inclusion of marginalized groups proved to be difficult; in some cases they could have been expecting short-term results and rewards, and in other cases they were either passive or distrustful, refusing to see the value of engaging in such a project.
7. Many of them went beyond what was "expected" from them. For example, one group sought to get legal status and started a new foundation. Another local

community got really motivated to improve its computer and oral presentation skills, and obtaining the volunteer services of a school teacher to give lessons. In at least two cases, the project helped citizens strengthen or scale up already existing local initiatives by adding value.

8. However, in our project, one farmer who was completely illiterate participated actively in discussions; in another group, the participants solicited the help of their sons and nephews to support them in the use of Information Communication Technologies (ICTs).

9. For additional information, visit https://www.ucl.ac.uk/excites.

10. https://daily.zooniverse.org/2014/09/16/citizen-science-in-dictionary/.

References

In the spirit of fair, open, and accessible science, we would recommend to those who cannot access some of the following content to use SciHub (the "Robin Hood of Science") and/or Library Genesis.

Agrawal Arun, and Clark C. Gibson. 1999. "Enchantment and Disenchantment: The Role of Community in Natural Resource Conservation." *World Development* 27 (4): 629–49.

Ashby Jaqueline A., Ann R. Braun, Teresa Gracia, María del Pilar Guerrero, Luis Alfredo Hernández, Carlos Arturo Quirós, and José Ignaicio Roa. 2001. *Investing in Farmers as Researchers: Experience with Local Agricultural Research Committees in Latin America.* Cali, Colombia: Centro Internacional de Agricultura Tropical.

Batson, C. Daniel, Nadia Ahmad, and Jo-Ann Tsang. 2002. "Four Motives for Community Involvement." *Journal of Social Issues* 58 (3): 429–45.

Brereton, Margot, Paul Roe, Ronald Schroeter, Anita Lee Hong. 2014. "Beyond Ethnography: Engagement and Reciprocity as Foundation for Design Research Out Here." In *Proceedings of the SIGCHI Conference on Human Factors in Computing Systems,* 1183–6 New York: ACM Press.

Chambers, Robert. 2006. "Participatory Mapping and Geographic Information Systems: Whose Map? Who is Empowered and Who is Disempowered? Who Gains and Who Loses?" *The Electronic Journal on Information Systems in Developing Countries* 25 (2): 1–11.

Cooke, Bill and Uma Kothari. 2001. *Participation: The New Tyranny?* London: Zed Books.

Comandulli, Carolina, Michalis Vitos, Gillian Conquest, Julia Altenbuchner, Matthias Stevens, Jerome Lewis, and Muki Haklay. 2016. "Ciência Cidadã Extrema: Uma Nova Abordagem. Monitoramento da conservação da biodiversidade." *Instituto Chico Mendes de Conservação da Biodiversidade* 6 (1): 34–47.

Deci, Edward, and Richard Ryan. 2000. "The 'What' and 'Why' Of Goal Pursuits: Human Needs and the Self-Determination of Behavior." *Psychological Inquiry* 11 (4): 227–68.

Dickinson, Janis L., Jennifer Shirk, and David Bonter. 2012. "The Current State of Citizen Science as a Tool for Ecological Research and Public Engagement." *Frontiers in Ecology and the Environment*, 10 (6): 291–97.

Finn, Janet. 1994. "The Promise of Participatory Research." *Journal of Progressive Human Services* 5: 25–42.

Ford, J. D., E. Stephenson, A. Cunsolo Willox, V. Edge, K. Farahbakhsh, C. Furgal, et al. 2016. "Community-based Adaptation Research in the Canadian Arctic." *Wiley Interdisciplinary Review of Climate Change* 7 (2): 175–91.

Kirkland, Rene A., Nancy J. Karlin, Megan Babkes Stellino, and Steven Pulos. 2011. "Basic Psychological Needs Satisfaction, Motivation, and Exercise in Older Adults." *Activities, Adaptation and Aging* 35: 181–96.

Mery, Gerardo, Pia Katila, Glenn Galloway, Rene I. Alfaro, Markku Kanninen, Maxim Lobovikov, and J. Varjo, eds. 2010. *Forests and Society – Responding to Global Drivers of Change. IUFRO World Series Volume* 25. Vienna: International Union of Forest Research Organizations. https://www.iufro .org/science/special/wfse/forests-society-global-drivers/.

Rotman Dana, Jenny Preece, Jennifer Hammock, Kezee Procita, Derek L. Hansen, Cynthia Parr, Darcy Lewis, and David W. Jacobs. 2012. "Dynamic Changes in Motivation in Collaborative Citizen-Science Projects." In *Proceedings of the ACM 2012 Conference on Computer Supported Cooperative Work*, 217–26. New York: ACM Press.

Ryan, Richard M., and Edward L. Deci. 2000. "Intrinsic and Extrinsic Motivations: Classic Definitions and New Directions." *Contemporary Educational Psychology* 25 (1): 54–67.

Tyler, Tom. 2010. *Why People Cooperate: The Role of Social Motivations.* Princeton, NJ: Princeton University Press.

CHAPTER 5

Contextualizing Openness: A Case Study in Water Quality Testing in Lebanon

Salma N. Talhouk, Rima Baalbaki, Serine Haydar, Wassim Kays, Sammy Kayed, Mahmoud Al-Hindi, and Najat A. Saliba

Abstract

Using participatory research methods, this Lebanon-based project engaged citizen scientist volunteers (predominantly women) to explore whether open and collaborative science could be used as an opportunity for environmental managemen and local development. Using data from a participatory mapping activity, fifty villages were selected that had identified "water quality" as a key area of concern. Local citizen scientists were then trained by the research team to conduct water-quality testing. After rounds of collecting water samples and analysis, researchers found that volunteers were more informed about local water issues, more likely to voice their concerns to political representatives, and, hence, take increased ownership over their community's health and well-being.

Introduction

This project sought to explore how a citizen science approach, i.e., opening up scientific inquiry to a broader public, could allow a more diverse group of people to participate in research and open exchange of scientific knowledge. Initial work by Buytaert et al. (2014) has suggested that through citizen science projects, individuals driven by an environmental concern can become part of a scientific process that allows them to generate data and help find answers to problems in an open knowledge-sharing environment. Similar work has highlighted that some citizen scientists see their involvement as a hobby driven by scientific curiosity (Cohn 2008; Buytaert et al. 2014), while others volunteer to learn about the environment where they live and to become directly involved in the planning of local environmental decisions that concern them, their families, or their community (Overdevest et al. 2004). The project team was interested in testing a citizen science approach to tackling water quality issues in Lebanon where, for the past fifty years, citizens have lived in a system plagued by war, political instability, and corruption, and incredible urban growth at the expense of the country's quality of water. Under such conditions, this project was interested in understanding how local remediation might be promoted by engaging citizen scientists in rapid water quality assessments.

This work built on research related to methodologies and ideas as to how to improve openness in the field of citizen science so as to benefit both scientists and citizens. For example, several works have looked at how knowledge sharing in citizen science contributes to the scientific literacy of citizen scientists who may engage in projects because of their personal interest or local crises (Conrad and Hilchey 2011; Fore et al. 2001; Silvertown 2009). The team wondered if by engaging in the scientific research process, citizens would also feel more empowered and responsible to speak up and take steps to improve their community water sources based on the data collected. Building on insights by Conrad and Hilchey (2011) and Fore et al. (2001), the team was interested in testing whether citizen scientists in Lebanon might develop a personal stake in the research through the processes of being trained, collecting data, and conducting scientific analyses. The project was also keen to better understand how scientists could benefit from citizen science through a partnership with citizens in an open communication process. Could such partnership(s) create a foundational trust between citizens and scientists, streamlining more open exchange? Sheppard and Terveen (2011) suggest that partnership can be strengthened and the quality of collected data can be improved through the design of basic charting interpretive tools, automation of advanced analyses, and generation of easily understood reports that allow volunteers to explore data themselves. These are broad principles that facilitate citizen involvement in scientific projects; however, they should not be prescriptive. Thus, the project team was interested to understand if a citizen science approach could be adapted to the citizens' interpretive capabilities and to the social and cultural dimensions of the research area. In addition to informing strategies for research application, understanding the social and cultural dynamics of the local area can better inform community engagement. To achieve these diverse outcomes, the project team determined that a multidisciplinary research team would be important. This diversity of disciplines from the science and social science fields could help to ensure that the focus was not only on scientific integrity, standardized methodology, and data validation, but also on training and engaging volunteers, and using appropriate technology to disseminate project data and results (Bonney et al. 2009; Silvertown 2009).

The team was interested in exploring a more participatory, "bottom-up" approach to the production of scientific knowledge to understand how it might enable a more equitable exchange where the interests of citizens and scientists are taken into consideration during the planning and the implementation phases. In countries with large numbers of poor individuals, marginalized communities, and depleted environments, citizen science appears to be potentially far more impactful than traditional science since knowledge generation in partnership with citizens can contribute directly to development. By engaging citizens through a more bottom-up approach, citizen science research appears to hold great potential to contribute to the development of the community by not only generating valid and important scientific outcomes but also by helping to serve as an effective awareness campaign. Through prolonged participation in researching and understanding an issue of local concern, citizens internalize the implications of the results and can formally and informally communicate those results to neighbours and other community members. We have previously introduced a methodological framework that highlights the crucial contribution of research to development in marginalized and politically unstable environments (Tawk and Talhouk in review at time of printing). The framework was tested in fifty Lebanese villages and produced community-generated data on local natural and cultural landmarks and contributed to local action initiatives.

In this project, the same open information-sharing framework between a university and a community was applied to assess the domestic water quality in a Lebanese village and to lay a strong groundwork for suitable solutions that may follow suit. Citizens tested the water quality of the main public and private wells feeding the village.

Two campaigns (August 2016 to September 2016 and November 2015 to February 2016) were conducted with each consisting of three sampling events followed by discussions. In this chapter, we shed light on the situational context of water quality in Lebanon at the village scale; we describe the methodology and process our citizen science project went through to engage and train community members; and we elaborate on stakeholders' responses and exchange. We close by proposing a revised framework that shows how citizen science may be used as a tool for community development and to help build a foundation for remediating local environmental problems.

University-Community Research and Development Framework: Lebanese Village Case Study

For the past fifty years, Lebanese citizens have lived in a system plagued by war, political instability, and corruption. As a result, urban growth has occurred at the expense of the country's natural resources, with water quality being a central issue affecting people's health and well-being. Discharge of raw, untreated sewage and industrial wastewater and dumping of solid waste in rivers, unregulated tapping into aquifers, and absence of storm water collection are examples of practices that have resulted in the deterioration of surface and groundwater quality. Under such conditions, is it possible to promote local remediation actions by engaging citizen scientists in rapid water quality assessments?

Aligning our research question with the local context was achieved by selecting a village community that expressed concerns about the quality of domestic water and placed cleaner water among the three top environmental priorities for local action. This was determined following an extensive participatory mapping process to help local communities identify natural and cultural landmarks in their towns and villages. This process involved the establishment of a local committee that assessed, mapped, and engaged in planning exercises around the cultural and natural landmarks of their village. In line with our participatory approach, the community members involved in the mapping activity were given equal standing and voice, and no member played a more significant role than another. During the consensus building and planning stage, the decision to make water quality a local priority was made collectively. The same work was conducted with more than fifty villages; however, this village was selected because its collective prioritization of water quality issues matched the priority of our team. The municipal council of the targeted community was then approached with the proposed citizen science project and the objective was explained. Upon receiving an expression of interest to participate from the local authorities, the project team secured Institutional Review Board (IRB) approval to ensure compliance with the university's ethics code. Unlike similar interdisciplinary research applying participatory research methods, there were no conflicting priorities or understandings on how to engage local people during the IRB process (Traynor et al. 2015; see also Chapter 10).

The municipality took charge of inviting the local community to the introductory seminar and introducing the research team to the owners of the private wells and the operators of the public wells. In assembling a group of citizen scientists, two citizens who were especially passionate about our work took the initiative to recruit participants. One of these citizens was instrumental in keeping the citizen scientists engaged in the training and water quality testing. Having enthusiastic local residents can be key to maintaining citizen engagement during training and fieldwork.

Discussions with various stakeholders allowed for an open exchange of information in relation to their interest in the proposed citizen science project. The municipality's main incentive to participate in this study was to verify the allegations of the Ministry of Health regarding one of the village water sources, which was officially reported as contaminated and not suitable for use. The research team was provided with a copy of the Ministerial decision. The private well owners were reluctant to participate because they were concerned about the lack of objectivity and reliability of tests performed by local residents. This concern was addressed by explaining the methodology, which consisted of blind sample testing and verification of the results in the university laboratories. The credibility of a university partner assisted in both quelling the concerns of private well owners and diffusing tensions among the many stakeholders surrounding the sensitive issue of local water quality. Also, municipal authorities appreciated being able to discuss their university partnership with other public officials. Without the formal role of a ground-based academic partner, the citizens may not have been able to adopt the right procedure for water quality testing, stakeholders may not have had confidence in results, and there may have been more tension in the community when the water quality results were made public. Citizen scientists wanted to know which water source they were testing and to have information about water quality in their homes. We explained the importance of objective analysis and blind samples and the need to assess all the water sources of the village.

Methodology

Participatory water quality sampling, as well as assessment and monitoring schemes for lakes, streams, rivers, catchments, and reservoirs have been implemented in several locations across the northern hemisphere (Au et al. 2000; Burgos et al. 2013; Latimore and Steen 2014; Overdevest et al. 2004; US EPA 2012), Australia (Nicholson et al. 2002), and, to a lesser extent, in a number of developing countries (Deutsch et al. 2005; Nare et al. 2006; Nare et al. 2011). Various water quality parameters were measured and collected during these campaigns using instruments with varying levels of complexity. Despite a number of challenges, such as funding, sustainability, reliability of data, demonstrable application of results, and the impact on water resource management decisions on both the local and national scale, the majority of these campaigns resulted in "synergistic outcomes" that included the advancement of freshwater science, public awareness of water resource challenges and concerns, increased levels of "citizen participation," and implementation of "science-based" protection/conservation projects at the local level (Burgos et al. 2013; Latimore and Steen 2014).

In our case study project, there was an agreement with the local residents and authorities that the main water sources in the village would be sampled for testing. The point source locations were shared by a multi-stakeholder consultation group, which included municipal representatives, a water authority representative, and private well owners. Water sampling was performed in coordination with this group and was based on the *Quick Guide to Drinking Water Sample Collection* published by the United States Environmental Protection Agency (EPA) in 2015 (US EPA 2015).

Simple field-testing kits and laboratory supplies for twelve water parameters were purchased or assembled by the team. The methods were selected because they utilized standardized procedures, were easy to use, had short assay durations, and were readily implementable in an improvised laboratory setting.

Volunteer Training

The solicitation for citizen scientist volunteers took place at the end of a public seminar in which we presented a background section (water as a resource, sources of water pollution, and water quality parameters), and a citizen science research section where we explained the need for regular testing and elaborated on the potential role that local residents can play as citizen scientists. The great majority of residents believed that all their water resources were contaminated, indicating that the village does not have a proper sewage system and that open wastewater dumping was compromising local water quality. One resident asked if we were there to provide filtration devices at the household level. Two residents believed that drinking contaminated water has no adverse impacts on health. Many of the citizen scientists were under the impression that testing the quality of water could be done with a simple probe; however, the training process shifted their perception by demonstrating the lengthy process of testing for only twelve core parameters.

Despite opening participation to all residents, only women were involved in the training workshops with the exception of one man. This could have been the result of a couple of factors. Two of the most enthusiastic citizens played a primary role in recruiting participants; they are women who have active roles as leaders in a local women's club. Another possibility may be due to the cultural context; women scientists were the key university figures during the project, and this may have influenced citizens' decisions to participate in the project. The university team did not enforce gender balance among the citizen scientists.

Training workshops were organized according to the local interpretive capabilities of citizen scientists. During these workshops, citizen scientists learned about water quality standards and testing procedures. Specifically, they learned that water quality depends on multiple parameters, of which we were testing only twelve; water must be tested regularly (preferably once every month); and different procedures exist for the testing of different parameters.

Fact sheets (see Figure 5.1) were designed for each parameter and distributed to the trainees. The fact sheets included an introduction about each parameter (what it is, why we care about it, the permissible level, and the measurement technique), a section detailing the testing methodology using illustrations, and a section about safety information.

Figure 5.1. One example of a fact sheet developed for the pH measurement in English (left) and Arabic (right).

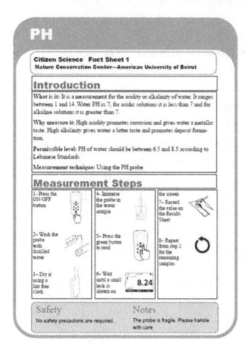

Citizen scientists followed procedures detailed in the fact sheets. Despite careful selection of scientific wording to cater to the interpretive capacity of laypersons and translation of the fact sheet into Arabic, in some instances, the citizen scientists did not quickly understand scientific jargon used in the fact sheet. For example, translation of pH to Arabic proved confusing to some citizens, but after some explanation and examples of quotidian acidic and basic household items, the citizens grasped the concept. Citizens were eager to start conducting the water quality tests, and we had to explain the importance of first understanding the fact sheet and procedure. Many of the citizen scientists expressed excitement over the water quality tests that changed colours and when tests yielded results that were well over the acceptable Lebanese standard for the parameter. When they were uncertain about the performance of some tests, they decided to repeat them. By the last testing workshops, the need for our intervention became minimal, and experienced participants were left to operate on their own. An interesting aspect worth reporting is peer teaching that took place. This is indeed an extension of rural Lebanese village practices where women often work together to prepare food preserves and dried herbs. Participants were paying attention to each other's work and correcting each other. For example, during the drop count titration, three women working together each kept her count and confirmed the number of drops at the end of the experimental tests. This cooperation between women also occurred during data collection; for example, when one person was taking a measurement, the other would fill out the results form. During laboratory-based water quality testing by experts, such cooperation is common. By teaming up to take measurements and record data, both citizen scientists and expert scientists can improve the efficiency of experimental work. Not all citizen scientists were able to participate in all testing workshops, and participants changed from one workshop to the other. Thus, we depended greatly on the regular participants to teach the procedures to newcomers.

Throughout the six full-day workshops conducted over a one-year period, we were able to observe different attitudes and interests among the citizen scientists. There were many community and cultural events, and turnover among the citizen scientists was relatively common. This was in part due to the open platform of the participatory approach and the ethics enumerated by the IRB where continual participant engagement was voluntary and not enforced. Some citizen scientists joined further along in the project and some left but the core group remained, which proved crucial for training incoming recruits. One theme that emerged from the citizen scientists was that some seemed interested in the technical aspects of the tests and not the outcomes, while others were interested to know more about the quality of each water sample and which sample corresponded to the water source that fed their house.

During every workshop session, participants were asked for their feedback, and revisions were made to the participation methodology. For example, instead of having consecutive testing sessions for water quality measurements, the sessions were scheduled simultaneously, each at a separate table, and additional sample cells were bought to eliminate the washing step between tests. This reduced the time of the workshop from four to two hours and allowed both the water collection and testing to be conducted on the same day. During the second campaign held between August and September of 2016 when children were on school break, some citizen scientists came with their daughters on weekdays. It is likely a cultural matter that mothers brought their daughters, and not their sons, to workshops that had turned into all-women activities. The workshops included, in addition to the chemical tests, two biological tests (total and fecal coliforms) and, as such, four working stations were formed: (1) PH, conductivity and turbidity; (2) hardness and alkalinity; (3) colorimetrics for nitrates, nitrites, ammonia, and sulfates; and (4) biological tests.

Data Validation

The purchased and assembled test kits were placed inside two plastic boxes on wheels for easy transport. The boxes were taken back and forth to the village for the workshops and to the laboratory to repeat the tests on the same samples. Data generated by the citizens were compared to the data generated in the laboratory using SPSS statistical software. Presentation of the findings will be published in a subsequent manuscript, which is in preparation.

Information Dissemination

Information dissemination occurred throughout the project period and was achieved in two ways, namely as side groups or one-on-one discussions and as formal public seminars where preliminary results were presented and possible actions discussed. During the course of the project, the university team served as a resource group for the citizen scientists and responded to all water quality–related enquiries whether they were directly related to the project or not. For example, we informed women concerned about the quality of purchased drinking water that the company owner should have a licence and follow the Ministry of Health's regulations. The team was also asked about local sources for water quality testing products in the case the citizen scientists wanted to retest their water in the future. The equipment used in this research project was expensive, particularly for the premade biological kits. The cheaper testing technology would have required an open flame, which the university team deemed dangerous. However, if the citizens want to retest water in the future or if we were to scale up this project, the cost could be substantially reduced if the materials are bought in bulk and the kits for the citizen scientists are manually prepared by the team of scientists. The project team worked closely with the municipality to have one of the wells tested by a certified laboratory as recommended of the Ministry of Health. Simple measures were also shared with women concerned with quality of water such as brushing teeth with drinking water, cleaning water storage tanks regularly, and using chemical disinfectants like chlorine tablets.

The results from the water samples collected during the campaign between November 2015 and February 2016 were shared in a public seminar attended by the majority of the citizen scientists, the private well owners, representatives from the municipality, representatives from a local school, and some concerned residents. Citizen scientists were acknowledged, and residents expressed pride that the study was done by community members from the village and that the presentation was all about the village and its local people. The results revealed that water in one of the private wells was contaminated. The owner, who was informed of these results ahead of time, apologized to the audience and promised to install a water treatment unit before distributing water again. To date, no water treatment measures have been implemented; however, the well owner is no longer selling water. He and all of the well owners were glad to have their water tested at no cost, and they held a discussion among themselves to see how they could go about treating the water coming from their wells to ensure quality. Surprisingly, none of the well owners expressed any regrets over their participation or animosity about the public exposure of contamination in their wells.

As a result of the first measurements and dissemination strategies, women became better informed about the water quality in the village and voiced their opinion and concerns to the local authorities. The local authorities listened attentively to the community members. One women expressed how "this is the health of my family that we are talking about." In response, the municipality sent water samples from a closed well that previously served the majority of the village to a certified governmental lab. This allowed the municipality to obtain permission from the Ministry of Health to reopen the well after it committed to install a water purification system. It is important to note that to date, water treatment systems have not been installed to remedy any of the polluted water sources. Well owners and operators seem unlikely to treat their contaminated water because of the associated costs and their current financial resources. Even though the citizen scientists did not expect the project to lead to full remediation of polluted water resources, they were disappointed to see that the authorities did not take immediate action to treat the community's water. The citizen scientists remain active via a WhatsApp messenger group and are still exploring opportunities to influence the water authorities and the municipality.

University-Community Knowledge Sharing, Research, and Development

It is important and possible for citizen science researchers to contribute to development, especially when they have a desire for social justice and an intrinsic motivation to work with marginalized groups (Gastrow et al. 2016). In fact, universities that support this scope of academic research and that are grounded in developmental issues are the ones that will drive societal renewal (Schieffer and Lessem 2014). The case study presented is an excellent model that illustrates how research and community development can be synergistic, as summarized in Figure 5.2. The model, which consists of four dimensions (tools for mapping, a trust building strategy, incentives to participate, and a participatory methodology), emphasizes the integrated roles of the university and the community in advancing research and development.

At the research level, the interactive approach helped to enhance the quality of participation by the citizen scientists, starting from their willingness to share information about the current status and location of wells, their active recruitment of committed residents, and their commitment to learn and comply with testing methodologies. Development was elucidated through several levels, including the establishment of a mobile water testing tool kit and the development of fact sheets about water quality parameters and testing procedures that can be readily replicated in other Lebanese communities.

Conclusion

Through this project, citizen scientists acquired expertise in water testing and were able to contribute to local decisions regarding water quality. As is often the case with traditional scientific findings harbouring implications for development, this citizen science project has yet to lead to the remediation of polluted water sources. It might be possible that with additional time for the project and an increase in

the financial resources of the private well owners and local authorities, the sources of pollution will be addressed and the local water
will be treated. However, before these local actions can be taken, a
foundation for this change had to be set through the local water quality issues being well understood and publicly exposed, in addition to
the mobilization of a core group of community members. The citizen
scientists engaged in our project gained greater understanding of
the quality of their community's water and the importance of their
direct involvement in assessing the current situation and setting the
groundwork for remediating local environmental problems.

**Figure 5.2. Summary illustration of how research and community
development can be synergistic.**

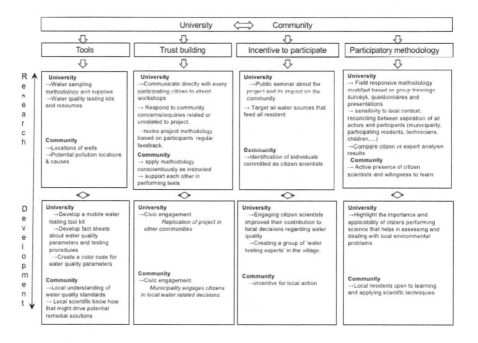

References

Au, J., P. Bagchi, B. Chen, Raul Martinez, Susan Dudley, and George J. Sorger. 2000. "Methodology for Public Monitoring of Total Coliforms, Escherichia Coli and Toxicity in Waterways by Canadian High School Students." *Journal of Environmental Management* 58 (3): 213–30. doi: 10.1006/jema.2000.0323.

Bonney, Rick, Caren B. Cooper, Janis Dickinson, Steve Kelling, Tina Phillips, Kenneth V. Rosenberg and Jennifer Shirk. 2009. "Citizen Science: A Developing Tool for Expanding Science Knowledge and Scientific Literacy." *BioScience* 59 (11): 977–84. doi: 10.1525/bio.2009.59.11.9.

Burgos, Anna, Rosaura Páez, Estela Carmona, and Hilda Rivas. 2013. "A Systems Approach to Modeling Community-Based Environmental Monitoring: A Case of Participatory Water Quality Monitoring in Rural Mexico." *Environmental Monitoring and Assessment* 185 (12): 10297–316. doi: 10.1007/s10661-013-3333-x.

Buytaert, Wouter, Zed Zulkafli, Sam Grainger, Luis Acosta, Tilashwork C. Alemie, Johan Bastiaensen, et al. 2014. "Citizen Science in Hydrology and Water Resources: Opportunities for Knowledge Generation, Ecosystem Service Management, and Sustainable Development." *Frontiers in Earth Science* 2: 1–21. doi: 10.3389/feart.2014.00026.

Cohn, Jeffrey P. 2008. "Citizen Science: Can Volunteers Do Real Research?" *BioScience* 58 (3): 192–97. doi: 10.1641/b580303

Conrad, Cathy C. and Krista G. Hilchey. 2011. "A Review of Citizen Science and Community-Based Environmental Monitoring: Issues and Opportunities." *Environmental Monitoring and Assessment* 176 (1–4): 273–91. doi: 10.1007/s10661-010-1582-5.

Deutsch, W., A. Busby, J. Orprecio, J. Bago-Labis, and E. Cequina. 2005. "Community-Based Hydrological and Water Quality Assessments in Mindanao, Philippines." In *Forests, Nature, and People in the Humid Tropics: Past, Present and Future Hydrological Research for Integrated Land and Water Management*, edited by Michael Bonell and L. A. Bruijnzeel, 134–50. Cambridge: Cambridge University Press.

Fore, Leska S., Kit Paulsen, and Kate O'Laughlin. 2001. "Assessing the Performance of Volunteers in Monitoring Streams." *Freshwater Biology* 46 (1): 109–23. doi: 10.1111/j.1365-2427.2001.00640.x.

Gastrow, Michael, Glenda Kruss, Maitseo Bolaane, and Timothy Esemu. 2016. "Borderline Innovation, Marginalized Communities: Universities and Inclusive Development in Ecologically Fragile Locations." *Innovation and Development* 7 (2): 211–26. doi: doi.org/10.1080/2157930X.2016.1200970.

Latimore, Jo A., and Paul J. Steen. 2014. "Integrating Freshwater Science and Local Management Through Volunteer Monitoring Partnerships: The Michigan Clean Water Corps." *Freshwater Science* 33 (6): 686–92. doi: 10.1086/676118.

Nare, Lerato, David Love, and Zvikomborero Hoko. 2006. "Involvement of Stakeholders in the Water Quality Monitoring and Surveillance System: The Case of Mzingwane Catchment, Zimbabwe." *Physics and Chemistry of the Earth, Parts A/B/C* 31 (15/16): 707–12. doi: 10.1016/j.pce.2006.08.037.

Nare, Lerato, John O. Odiyo, Joseph Francis, and Natasha Potgieter. 2011. "Framework for Effective Community Participation in Water Quality Management in Luvuvhu Catchment of South Africa." *Physics and Chemistry of the Earth, Parts A/B/C* 36 (14/15): 1063–70. doi: 10.1016/j.pce.2011.08.006.

Nicholson, Emily, Jane Ryan, and D. Hodgkin. 2002. "Community Data— Where Does the Value Lie? Assessing Confidence Limits of Community Collected Water Quality Data." *Water Sci Technol* 45 (11): 193–200.

Overdevest, Christine, Cailin Huyck Orr, and Kristine Stepenuck. 2004. "Volunteer Stream Monitoring and Local Participation in Natural Resource Issues." *Human Ecology Review* 11 (2): 177–85.

Schieffer, Alexander, and Ronnie Lessem. 2014. "The Integral University: Holistic Development of Individuals, Communities, Organisations and Societies." *PROSPECTS* 44 (4): 607–26. doi: 10.1007/s11125-014-9324-z.

Sheppard, S. Andrew, and Loren G. Terveen. 2011. "Quality Is a Verb: The Operationalization of Data Quality in a Citizen Science Community." In *Proceedings of the 7th International Symposium on Wikis and Open Collaboration*, 29–38. New York: ACM Press, doi: 10.1145/2038558.2038565.

Silvertown, Jonathan. 2009. "A New Dawn for Citizen Science." *Trends in Ecology & Evolution* 24 (9): 467–71. doi: 10.1016/j.tree.2009.03.017.

Tawk, L. Y. and Salma N. Talhouk. Forthcoming. "A Developmental Framework For Cultural Ecosystem Services Research in Conflict and Marginalized Areas: Lessons Learned From Lebanon." *Ecosystem Services*.

US EPA (United States Environmental Protection Agency). 2012. "Monitoring and Assessing Water Quality - Volunteer Monitoring." https://archive.epa.gov/water/archive/web/html/index-18.html.

———. 2015. *Quick Guide to Drinking Water Sample Collection*. 2nd ed. Golden, CO: US EPA. https://www.epa.gov/sites/production/files/2015-11/documents/drinking_water_sample_collection.pdf.

SECTION 2

GOVERNING OPEN SCIENCE

INTRODUCTION

Cameron Neylon

At the core of all projects—whether research, open or closed, collaborative or internal—is the question of governance. For many projects governance is implicit, determined by unwritten assumptions and cultural practices. Indeed, raising the idea of making those implicit assumptions explicit can be seen not just as a threat, but also as an accusation of bad faith. Why should we require formal rules or processes when it is obvious to all people of good faith how we should behave?

Such universal agreement is, of course, rare, if not non-existent, in practice. However, certain approaches to work, and to research specifically, bring those differences to the surface more clearly. Among these, open and collaborative practices are very likely to raise issues. If the goal of open and collaborative science is to include a more diverse range of actors, to give greater agency to those who have lacked it, and to gain from perspectives that are frequently excluded, then we must expect such projects to raise issues of governance. It is not that they are unique in doing so, but that in placing the value of difference at the centre, they bring the consequences of difference to the surface.

At the same time, open and collaborative projects often suffer from a particular version of the general blindness to governance challenges. Because they are frequently driven by values, often including labour that is volunteered or above the minimum required, it is easy to leave the differences in those values unexamined. Ironically, open and collaborative projects frequently create their own form of exclusion, identifying those who "don't contribute" or "don't understand" as outsiders, while reinforcing an assumption of homogeneous values and motivations among insiders. Even when projects explicitly focus on the inclusion of different communities, there can be surprises when these different internal groups are found to have differing motivations.[1]

Governance, at its best, formalizes that which needs to be made formal and leaves space and flexibility for customary practice. In general, the need for formal rule-making increases as the size and diversity of a community increases. The provision of governance institutions is a collective action problem (Ostrom 1990), and as with all collective action problems, it is best solved with smaller and more homogeneous groups (Olson 1974). But a group with overly formalized rules is likely to be rigid and unwelcoming and, therefore, unlikely to grow. The best governance systems will surface issues in a context in which the systems to address them can be developed, using difference to advantage. Similarly, the best open and collaborative research projects will harness diversity to create value. A similar tension exists between internal efficiency of communication and common understanding, and the capacity to gain value from different perspectives. That is, the governance problems for projects are similar to the challenges of building the most effective open and collaborative projects.

The three chapters in this section tackle these questions from a range of perspectives. They raise questions of trust, of the necessity of formalized agreements, of how new contributors can be encouraged, and of how common languages can be built to support collaboration. At the centre of these are issues of trust and of control: trust in the actions of collaborators, trust in the institutions that help manage the interactions, and control over assets—over how they are shared and used. Contributors need to be in control of the process through which they begin to trust the project, its members, and the systems within which they work. As that trust develops, control over "their" assets becomes less necessary, but a sense of contributing to the control over the project, of there being a controlled process, becomes crucial.

Building Trust: Prior Agreement and Allowing Control

By definition, collaborations bring different groups together, but, even within groups, the question of trust is critical. In their contribution, Maurice Bolo, Victor Awino, and Dorine Odongo tell an all-too-common story: A failure to agree on arrangements for management and control over outputs in advance leads to conflict, which, even when resolved, can leave residual resentment (see Chapter 8). The desire to avoid discussing, in advance, who has control over the opportunities that arise from research comes from two places. The first, and most frequent, is a stated belief that "everyone is on the same side;" the second is the reality that mechanisms and systems of agreement often require giving up control.

The same chapter also presents three distinct case studies that illustrate differing consequences arising from a lack of prior agreements. In Case Study 1, the lack of such agreements led an industrial partner to negotiate with the funder to take control of assets from the other partners. In Case Study 2, project participants answered the question of whether or not a property claim would be made with "we shall wait and see," illustrating a lack of trust in the project as a whole and a consequent wish to retain control until forced to give it up. Unfortunately, as the case study notes, at that point it is often too late to resolve issues without significant conflict. Finally, in Case Study 3, one partner was said to have "run with our knowledge," taking collective findings and seeking individual advantage without reference to the partners. In these situations, prior agreement is crucial both to ensure control by the collaboration, but also to build trust in the collaboration as an entity in the longer term. The three case studies suggest a systemic lack of faith between the partners within collaborations in Kenyan applied research.

In their contribution, Dora Canhos and collaborators discuss a different, albeit related, experience. With the goal of creating a shared infrastructure for sharing biodiversity data—Brazil's Virtual Herbarium (BVH)—they started with the assumption that all data would be publicly and fully shared (see Chapter 6). In the authors' words, "in the name of openness, in order to participate, all data had to be shared." While the various herbaria that contributed data were used to sharing in an informal way with other scientists, this shift to public sharing was new and raised concerns. Sharing beyond the community meant that data were out of their control—and they did not yet trust in the system.

Contributions to the project were slow until mechanisms were provided that allowed the participating herbaria to hold back data that they deemed sensitive. In particular for endangered or otherwise sensitive species, the data might be held back and kept under their control. In combination with the provision of new information and support that the central infrastructure of the BVH could provide, this has grown over time, increasing engagement from data contributors and ultimately building a community with agreed-upon procedures and rules. While the formative document for the BVH is a non-binding memorandum of understanding, it nonetheless sets out important expectations and, therefore, builds community.

Rule Making and Community Building

The non-binding nature of the BVH memorandum is sufficient, despite the involvement of over a hundred institutions, because the risks are relatively small. The control granted to data contributors, along with the fact that those contributors do not directly contribute funding to the BVH, makes a statement of principles sufficient. This would likely change if the constituent institutions were contributing funds to BVH or if there was an external mandate to contribute all of their data. It may change as the services provided by BVH become more important to the operations of the contributing herbaria.

In Chapter 7, Maurice McNaughton and Lila Rao-Graham note the importance of a pre-existing agency, the Caribbean Disaster Emergency Management Agency (CDEMA), in their work on building a shared vocabulary to support disaster management coordination in the Caribbean. This transnational agency coordinates efforts, including information sharing, across a set of countries where shared information and government transparency are not generally priorities.

The context of disaster management is unusual in two ways. First, environmental disasters, particularly hurricanes, do not respect national boundaries and coordination therefore has high value. Second, effective coordination in the context of these disasters, with communication systems often limited or disabled, needs to be governed by well understood and shared rules. McNaughton and

Rao-Graham describe how disaster management is an area where the benefits of trust building and shared procedures are particularly clear.

Arguably, the lack of rule making in the context of Kenyan research exploitation described by Bolo et al. is preventing the kind of community building that would support the policy goals of the Kenyan government and universities that they describe. The high-level aspirations found in the Kenyan constitution and university policies appear not to be implemented in practice. In part, this can be ascribed to a lack of community and culture to mediate the translation of high-level aspirations into practice. This, in turn, reflects the apparent lack of trust in community norms and culture by individual researchers.

The work of the BVH described by Canhos et al. can also be seen in the light of community building. Some compromises may be made to encourage engagement by a wider range of contributors, that engagement being driven by well-understood and specified rules about how the contributed data are used. An important question to ask is whether mechanisms are in place to allow the community to change those rules if it so desired. Achieving unanimity on a contentious subject could be challenging, which could necessitate a more formalized approach.

The interaction between informal norms, or culture, formal rule making, and trust within communities is complex and highly dependent on context and history. What is common to all three of these studies, presented in the chapters that comprise this section, is that an understanding of the complexity of the relationship is key to the success of projects.

Building Trusted Institutions

All three studies can be understood as seeking to build institutions that support communities. Elinor Ostrom defines the term institution to cover "the prescriptions that humans use to organize all forms of repetitive and structured interactions" (Ostrom 2005, 3). This is most explicit in the case of BVH, where a technical infrastructure is being built, an institution that manages the interactions of the data contributors. The Open Knowledge Broker, as well as its underpinning vocabulary, can also be seen as an institution in its goals of supporting interactions between government and civil society agencies through a structured set of information exchange mechanisms.

On the surface, the case studies from Kenya in Chapter 8 may seem less focused on institution building. However, the implicit goal of their work is to identify an optimal set of prescriptions that researchers and industrial partners can use to organize a specific form of "repetitive and structured interaction." It is the absence of institutions in the Kenyan context to support the management of apparently conflicting demands of intellectual protection and exploitation, and open knowledge in support of development that Bolo and Awino criticize.

Functional institutions need to be trusted to work effectively. Trust, in turn, is a characteristic of communities. It could be argued that an institution becomes functional, in the complex sense of the term as dissected by Star and Ruhleder (1996), when there is sufficient trust from a defined community so they are happy to give up some forms of control in exchange for the shared benefits that the institution brings. That trust, as we have previously argued (Bilder, Lin, and Neylon 2015), requires more than just functional governance, but also economic and financial sustainability, as well as understood measures to deal with catastrophic change. Governance is not sufficient, although it is necessary. More than that, the need for governance changes as a community, project, or institution evolves. What is necessary to create trust at the beginning may be very different from what sustains trust in a mature project.

The Governance of Open and Collaborative Projects: The Value of Explicit Values

At the opening of this introduction, I noted a very common problem: the avoidance of discussions of governance. In contrast to the oft-given reason "but we all trust each other," I attributed this to a lack of trust, and trust that needs to be built up over time. Further, I argued that open and collaborative projects are both particularly prone to the challenges of bringing different perspectives together and, through an assumption of shared values under the banner of "open," also have a tendency to create differing forms of exclusion. In this final section, I argue that a shared institution of explicitly expressed values can mitigate these risks and strengthen the best aspects of open approaches.

The challenge that open and collaborative projects face is that they explicitly seek to bring together differing perspectives and, therefore, differing practices and cultures. It is this diversity that is thought to create the unique value of these approaches. At the same time, the values that underpin the desire to bring different communities together are often assumed and unexamined. In the Open Science Manifesto (OCSDNet 2017), the Open and Collaborative Science in Development Network argues in favour of making a set of seven core values explicit (see Chapter 2 of this volume).

These core values are challenging for many who claim the label of Open Science. There is much argument as to whether Open Science includes a commitment to diversity and inclusion. My argument is that those who do not share the values of diversity and inclusion will fall foul of the trust issue as they seek to build open and collaborative projects. Similarly, their lack of critical examination of the shared values they seek for their project will not only lead to conflict, but also to a failure to realize the opportunities that open and collaborative approaches bring.

In this sense, an explicit and shared set of values, as well as community processes that critically examine their application in practice, can be seen as a shared governance institution. Focusing on values allows flexibility for contextual rule making, but explicitly surfaces the issues around which formal rule making may be required. Those rules, in turn, may be refined into institutional form as rules-in-use as described by the Institutional Analysis and Design framework (Ostrom 2005). More than that, this evolving complex of explicit values and rules-in-use may ultimately provide a platform, an infrastructure, which enables like-minded people to come together rapidly to build open and collaborative projects.

Governance is a challenge, and the challenge changes as projects scale up. Ultimately, the challenge is one of building trust, and an explicit discussion of values is a strong place to start when building trust. The application of values as rules-in-use will always need to be contextual. Each rule implies a giving up of control by participants as they agree to be governed. That process in turn involves trust. In the end, governance is a question of privileging positive freedoms, the "freedom to," over negative freedoms, or "freedom from" (Holbrook 2015). The purpose of collaboration is to bring many together so that we may go far. To do that, we must have sufficient trust in the systems that help us work together so that we are willing to give up some elements of control.

Notes

1. For example, GalaxyZoo offers authorship to its contributors.

References

Bilder, Geoffrey, Jennifer Lin, and Cameron Neylon. 2015. "Principles for Open Scholarly Infrastructures-V1." *Figshare*. February 23, 2015. http://dx.doi.org/10.6084/M9.FIGSHARE.1314859.

Holbrook, Peter. 2015. *English Renaissance Tragedy: Ideas of Freedom*. The Arden Shakespeare. London: Bloomsbury Academic.

Olson, Mancur. 1974. *The Logic of Collective Action: Public Goods and the Theory of Groups*. Rev. ed. Princeton, NJ: Harvard University Press.

Open and Collaborative Science in Development Network (OCSDNet). 2017. "Open Science Manifesto." *OCSDNET blog*. May 28, 2017. https://ocsdnet.org/manifesto/open-science-manifesto/.

Ostrom, Elinor. 1990. *Governing the Commons: The Evolution of Institutions for Collective Action*. Cambridge: Cambridge University Press.

——. 2005. *Understanding Institutional Diversity*. Princeton, NJ: Princeton University Press.

Star, Susan Leigh, and Karen Ruhleder. 1996. "Steps Toward an Ecology of Infrastructure: Design and Access for Large Information Spaces." *Information Systems Research* 7 (1): 111–34.

CHAPTER 6

Brazil's Virtual Herbarium, an Infrastructure for Open Science

Dora Ann Lange Canhos, Sidnei de Souza,
Vanderlei Perez Canhos, and Leonor Costa Maia

Abstract

Currently, a large digital infrastructure project known as the Virtual Herbarium allows for small and large biological collections to compile and share data for increased academic and public access to Brazilian botany records. This project sought to understand who is using these data and for what purposes, as well as to understand the institutional benefits of data sharing. The project reveals many of the benefits and complexities of scientific collaboration across institutions and between disciplines while revealing the importance of building Open Science infrastructures in participatory ways.

Introduction

The evolution of information and communication technology is changing not only the way knowledge is produced, but also the way it is communicated (Gibbons et al. 1994). Before mass education, scientists were viewed as the holders of knowledge, and scientific communication was largely restricted to the scientific community. Today, many scientific developments aim to solve specific problems involving specialists from different fields of knowledge, within different cultures, and working in different countries. Communicating science and knowledge must reach out to all members of the community of specialists who necessarily must be part of the process (Hobsbawm 1994). Therefore, the dissemination of results is not sufficient in itself; the process must also be documented and accessible. In many fields, science is also an object of public interest and subject to public discussion; hence its particular vocabulary is absorbed into and becomes a vernacular with a greater dissemination of scientific data and information to society (Nowotny et al. 2001). The landscape of communication and information technologies continues to evolve and, consequently, there is a growing demand for online, dynamic, real-time, and two-way information and communication systems that accompany a process throughout its lifetime and are available to different users.

Biodiversity and Sustainable Development

Brazil's constitution of 1988 and its national environmental policy indicate that it is the duty of the State and the citizens' right to have access to environmental information. The development of an e-infrastructure focused on biodiversity data, such as Brazil's Virtual Herbarium (HVFF, an initialism for Herbário Virtual da Flora e dos Fungos), not only contributes to the advancement of science, but also helps guarantee our constitutional right to environmental information.

Problems such as poverty, hunger, inequity, environmental degradation, genetic erosion, lack of access to water, and climate change, among many others, call for a new international agenda and framework for cooperation. The Convention of Biological Diversity (CBD), opened for signature in 1992, was inspired by the concept of sustainable development,[1] focusing on biodiversity conservation, sustainable use, and benefit sharing. The CBD indicates that all parties "shall facilitate the exchange of information, from all publicly available sources, relevant to the conservation and sustainable use of biological diversity" and that "such exchange of information shall include exchange of results of technical, scientific and socio-economic research" (United Nations Environment Programme 1992). The Millennium Development Goals signed by one hundred and ninety-one national states in 2000 sought international commitment to development and the elimination of poverty and hunger in the world, including goals concerning environmental sustainability and global partnership for development. The Strategic Plan for Biodiversity

2011–2020, including Aichi Biodiversity Targets (Convention on Biological Diversity 2010), indicates the requirement of biodiversity data to monitor and achieve the targets.

A country's commitment to sustainable development, ensuring economic growth that is socially just and environmentally sustainable (Sachs 2015) is fundamental and its practice must occur at all levels, from local to global. If this is to be accomplished, data, information, and knowledge must also be organized and shared at all levels.

A Platform for Open Science

The increase of knowledge on Brazil's biodiversity, associated with scientific advances to understand the evolutionary processes that generate and maintain this diversity, is fundamental to the sustainable use of this natural capital. The HVFF was launched in December 2008 as one of Brazil's National Institutes of Science and Technology (INCT— *Institutos Nacionais de Ciência e Tecnologia*) to document, store, disseminate, and increase the knowledge base on the diversity of plants and fungi of Brazil.

Brazil's INCT program aims to mobilize and gather the best research groups to participate in activities of high scientific impact and in frontier and strategic areas to solve great national problems. The ultimate goal is to form national and international scientific cooperation networks. Biodiversity and sustainable development are strategic areas of the program.

In the last decades, few large investments were made in developing cyber infrastructures to support research (Barjak et al. 2010). An example in Brazil is the National Education and Research Network (RNP, *Rede Nacional de Ensino e Pesquisa*) that provides connectivity services to the academic community based on internet technology. However, engineering breakthroughs alone are not enough to achieve the outcomes envisaged for the undertaking of Open Science and other global collaborative activities supported by cyber infrastructures. If these are to be achieved, it will more likely be the result of a nexus of interrelated social, legal, and technical transformations (David 2006).

HVFF undoubtedly benefited from the advancements of RNP and also of *species*Link, the digital infrastructure used as its information base.[2] Both are fundamental to its success. However, its capacity to integrate institutions and people as a network, with different roles but with common aims, is what makes the difference.

The HVFF project began with twenty-five national herbaria and two herbaria from abroad, repatriating data of samples collected in Brazil. These herbaria, together with another sixteen herbaria that were part of *species*Link, shared forty-eight data sets providing about 1.8 million data records online. Currently, HVFF, with one hundred and six associated national herbaria, twenty-five herbaria from abroad, and twenty other herbaria that are not associated with the project but that share their data through *species*Link, integrates and openly shares over 5.5 million data records from one hundred and ninety-one datasets and more than 1.4 million images (see Figure 6.1).

Figure 6.1. Number of datasets and geographic location of those herbaria which provide data to Brazil's Virtual Herbarium in December 2008 and March 2017

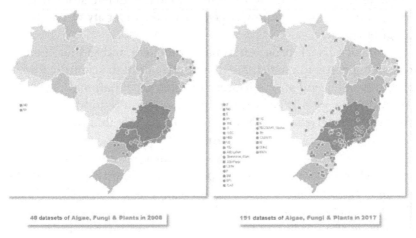

Source: *species*Link, 2017. *http://inct.splink.org.br/showNetwork*.

In addition to the presence of at least one herbarium in every Brazilian state, it is also important to acknowledge that almost ninety-five percent of national herbaria associated with HVFF are also associated with graduate programs. These circumstances, together with the easy access to online data and tools, are affecting the development of research and education in Brazil.

Strategies That Contributed to Openness

During *species*Link's early stages of development, biological collections had to openly share all data available in order to participate and receive funds. There were no mechanisms in place to hold back data considered sensitive or confidential. Therefore, in the name of openness, all data had to be shared. Sharing data within their own community was a normal practice among biological collections, but making data available online, to anyone interested without knowing who was accessing it and for what purpose, meant an enormous cultural change. Participation grew when mechanisms were built to ensure that data providers could easily hold back sensitive data.

Lessons Learned

A) *Data policy, including decisions as to what data can be shared openly, must be carried out at the data provider's end.*
The digital infrastructure adopted a general policy (CC BY-NC-SA 3.0[3]), and all data that are shared must follow the specific licence. Another important feature refers to expertise in informatics. Since the beginning, it was clear that most biological collections had very little expertise and inadequate infrastructures concerning informatics and, in most cases, connectivity. Therefore, the strategy was to adopt a simple architecture at the data provider's end and reduce demands, trying not to alter the collection's routine.

B) *The complexity of the network in informatics must lie at the digital infrastructure's end.*
The use of internationally accepted data standards and communication protocols was fundamental. Development of *species*Link began in collaboration with SpeciesAnalyst, a network developed at Kansas University in the US GBIF, the Global Biodiversity Information Facility, was also just beginning. All these initiatives came together and defined a common data model (DarwinCore) and a protocol (DiGIR, Distributed Generic Information Retrieval). Adopting internationally accepted standards and protocols enabled the integration of data from other networks, thus facilitating data repatriation.

C) The use of internationally accepted standards and protocols is essential.

The BVH project began with existing infrastructures, developed by the *Centro de Referência em Informação Ambiental* (CRIA), responsible for the development and maintenance of the *species*Link network. RNP was responsible for the backbone of the national academic network. Members of HVFF's steering committee are members of the Brazilian Botanical Society (SBB — *Sociedade Botânica do Brasil*) and its network of Brazilian Herbaria (*Rede Brasileira de Herbários*). These three initiatives, CRIA, RNP, SBB, and, evidently, the botanical community, are the pillars of this project, which would not have progressed if it disregarded existing initiatives.

D) When developing a digital infrastructure, focus on establishing strategic alliances with successful initiatives.

In addition to the data, many tools were developed in close partnership with the herbaria and the user community. The search interface[4] was largely enhanced, allowing users to produce maps, charts, and inventories with the results from their search. It also enabled users to compare images and to produce catalogues on the fly. An annotation system was developed to provide users with the means to help curators improve the quality of the data. Various tools were also developed to help curators to find inconsistencies and errors, thus improving the quality of their data. Reports providing all suspect or inconsistent records are available online[5] for both curators and users to attest to the quality of the data. These tools were greatly enhanced due to the close proximity to the herbaria. Beside raising data quality and increasing usage, these tools represent an incentive to data providers, a stimulus to participate.

E) Interact with data providers and users to develop necessary and meaningful tools, which are more effective and help motivate participation.

Governance

The network involves ninety-six national institutions and integrates data from herbaria belonging to another twenty-one institutions from abroad. There is a non-binding memorandum of understanding between data providers, the project leader, and CRIA, the institution responsible for the digital infrastructure. The initiative is project-based, meaning that it has limited resources and, in theory, a limited lifespan. Even with these limitations, governance is essential. Important features of good governance include participation, planning, accountability, and transparency.

Participation

HVFF is led by a project manager, a steering committee (researchers from the botanical community), and coordinators of specific topics (taxonomy, human resources, articulation, products, and online information systems). Face-to-face meetings are held at least once a year, and meetings with curators of participating herbaria are also organized annually during the National Congress of Botany. All other communication is carried out through the internet, websites, blogs, and emails.

Planning

The Steering Committee and coordinators use data from the digital infrastructure to set out specific strategies. As an example, one of the programs involves sending specialists to herbaria to identify material and offer courses. To organize this, the herbaria were asked to publish their unidentified material online, and a search interface was developed to enable searching for these records. This made it possible for the steering committee to select herbaria with the greatest need and match them with specialists with the necessary expertise. As a result, over seventy thousand specimens were examined and identified. Specialists from the network were also encouraged to visit herbaria to confirm identifications and update nomenclature, thus improving data quality. Courses on taxonomy and nomenclature were offered to graduate students, and courses on herbarium management and data quality were offered to curators and technical staff.

To establish priorities, a system called *Lacunas*[6] (Canhos et al. 2014) was developed to identify taxonomic and geographic information gaps in the data system. Through this tool, the steering committee prioritizes taxonomic groups for digitization and identifies understudied groups. Curators also use this information to develop strategies to guide their fieldwork.

A workflow to model species' distribution based on their ecological niche called BioGeo or Biogeography of Flora and Fungi of Brazil (*Biogeografia da Flora e Fungos do Brasil*)[7] was also developed. Through this system, volunteer specialists produce and publish geographic distribution models that are openly shared online and are used to help guide fieldwork and improve data quality. Over eight percent of the species listed in Brazil's list of plants and fungi have online geographic distribution models accessible to all who are interested.

As part of OCSDnet (Open and Collaborative Science in Development Network), together with HVFF, resources were available to:

1. involve curators in a participative SWOT analysis (Strengths, Weaknesses, Opportunities, and Threats);
2. develop metrics to measure data usage;
3. carry out a survey on who is using the system; and
4. analyze the motivation for voluntary participation of specialists.

SWOT Analysis

The SWOT analysis began with a questionnaire sent and answered by email. Preliminary results were presented and further discussed at a face-to-face meeting. Most curators indicated important benefits through open online data sharing, such as greater institutional recognition, increased involvement in graduate courses, more visits to the herbaria, growth in their holdings, and an increase in the number of grants. The most significant weakness identified in the analysis was a lack of staff, which cannot be addressed by the project. A significant opportunity is the importance of botany to sustainable development, and the most significant threat is the fact that HVFF is still project-based, which means that its future is uncertain.

The social network, strengthened through HVFF, has increased interaction between curators and technicians from different institutions, which, in turn, has led to a change in the mindset of the professionals involved, who now feel valued and a part of HVFF's achievements. The increased geographic coverage of the network, with the participation of small herbaria, is a very important asset, as many of these are regional collections whose holdings are underrepresented in other collections. Participation in HVFF promoted increased collaboration with students and researchers from other courses and institutions, as well as more visits from foreign researchers.

Usage Metrics

In terms of recognition, the metrics employed for measuring usage represented a problem for this project. Normal metrics of logins to the system include the number of hits, visits, pages, and bandwidth used. The country, state, or municipality from which users originate are normally based on the computers' IP (Internet Protocol) addresses. These metrics mean very little to this community and do not show the true impact of the Virtual Herbarium. As CRIA keeps a log of all searches within *species*Link's search interface, it is possible to indicate how many data records were actually used. Users carry out a specific search, and the system retrieves records that meet the search criteria. These records can then be listed, viewed in maps or charts, or downloaded. Records that are viewed are considered "used." The data for this indicator go back to October 2012, when the new search interface was launched.[8] The results reflecting the status in 2017 are presented below (Figure 6.2).

Figure 6.2. Number of data records used in Brazil's Virtual Herbarium between October 2012 and March 2017

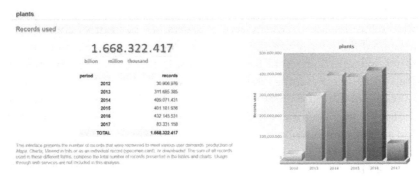

Source: *species*Link, 2017.

Between October 2012 and March 2017, about 1.7 billion plant and fungus records were used. The years 2014 and 2015 averaged a little over 400 million records, and 2016 showed over 432 million records, representing an average of 1.2 million records used per day.

As an example of the impact of this usage, a small but geographically specific herbarium in the State of Tocantins (HUTO) with six thousand records online can now show that over six hundred and twenty-seven thousand records were used in 2016. This represents one hundred and four times its online holdings and shows the impact of online data sharing.

Usage

Another important component of the OCSDnet studies was to identify the users' profile and to learn more about the purpose of using species occurrence data. When a digital infrastructure is developed, having a target user can help it to be more effective. Rather than building an infrastructure for a generic user, having a target user in mind enables identification of the needs and demands for the ideal target user, which can include thinking about what language(s) to use as well as how the data should be provided to be both useful and usable. Some of the target users in our case were researchers, graduate teachers, students, and Brazilian policy makers. Even if users whose profile or affiliation we had not anticipated go on to use and benefit from the open data, starting out with a target audience helps the development team keep a concrete scope and focus.

The survey directed to all *species*Link users is available online and dynamically shows the result when a form is submitted.[9] Based on six hundred and twenty-five answers, 43 percent of those who answered the survey use the system for their research, 20 percent for education, and 36 percent for other uses. In research, data are mostly used in taxonomy and systematics, biogeography, conservation, and ecology. For education, use is focused primarily in botany, ecology, and zoology. Other uses include data for species lists, fieldwork planning, Red Lists, environmental impact studies, environmental management, and public policies.

As for the users' profile: ninety-four percent are residents of Brazil, fifty percent have a doctorate degree or are PhD students, twenty-eight percent have a master's degree or are master's students, eighteen percent have higher education, and two percent have a high school education. As to their institution, 74 percent come from universities and research institutes, twelve percent from governmental institutes, six percent from the private sector, and four percent from schools and NGOs. This survey shows that, in the main, we are reaching out to our target users, although there are ten percent from sectors that we had not considered—these being the private sector, schools, and NGOs mentioned above.

Crowd Sourcing

Beyond identifying interactions with target users, the study carried out within the OCSDnet project framework also facilitated an identification of the motivations that led various users to contribute to the project. This was particularly useful in terms of evaluating the crowdsourcing component of the BVH—crowdsourcing understood as a term used for collaborative contributions such as voluntary services or ideas through the internet. HVFF developed two tools to use the expertise available to improve data quality (an annotation system) and to produce species geographic distribution models (BioGeo).

The main motivation for specialists to identify material and correct data through the annotation system was to contribute to improved data quality for their own research. The same applies to the species' geographic distribution model. Again, specialists were motivated mainly in order to use the system for their own research and in planning new collecting efforts.

Accountability and Transparency

A number of indicators were developed and are available online for all interested. These include indicators of the network, such as movement of data, geographic distribution of data and herbaria, and updating of the indexes.[10] Such indices are also set up for each herbarium and enable evaluations of any participating herbarium. Data quality indicators are also online for individual herbaria[11] and for the network, in this case, through the search interface.[12]

Conclusion

The necessary evolution of scientific communication to include different publics is especially true for botany and mycology and their importance to sustainable development. Challenges range from local to global, and "openness" is vital at all levels. However, there are many hurdles to overcome. Evaluation systems in universities and research centres are mostly based on individual metrics, though working as a team is essential. Networking and providing significant scientific services such as publishing and curating data are normally not valued, even when the availability of quality data is the basis for the advancement of science and for policy and decision-making processes. Publishing in journals of great international impact is what counts, even for developing countries, and this reduces the importance of local journals in local languages with a focus on local problems. Beside *global digital infrastructures* that integrate data worldwide, it is important to develop *local digital infrastructures*. Our experience indicates that the organization and dissemination of one's own data increases the capacity to use these types of data and represents the basis of the development of a true network.

Many funding agencies request that project proposals include strategies for managing and sharing data online. This is an important step, but it is not sufficient. For users to be able to rely on information systems, it is crucial for them to operate with uninterrupted, long-term funding, and these agencies operate through project-based strategies. For data that are permanent and must be preserved and offered over time, a digital infrastructure must be in place and must provide services to projects that produce such data. Digital infrastructures require long-term maintenance and constant development, continuous and dynamic evaluation and planning, and efficient governance models to assure continuity of the network and its services (Canhos et al. 2015).

HVFF's continuous success depends on consolidating the social network established and its digital infrastructure as a platform for Open Science to boost frontier developments in taxonomy, biogeography, conservation, ecology, and biodiversity informatics. It also depends on stable, long-term funding and the establishment of a governance model that is able to maintain its identity as a collaborative network.

Notes

1. Sustainable development is development that meets the needs of the present without compromising the ability of future generations to meet their own needs (WCED 1987).
2. See http://splink.cria.org.br for more information.
3. Creative Commons licence: Attribution (BY), Non-commercial (NC), Share-alike (SA).
4. See http://inct.splink.org.br for more information.
5. Select a dataset to see a report at http://splink.cria.org.br/dc. (Nb.: The page displays in Portuguese with an English language option.)
6. See http://lacunas.inct.florabrasil.net for more information.
7. See http://biogeo.inct.florabrasil.net (only available in Portuguese).
8. See http://www.splink.org.br/showUsage.
9. See http://www.splink.org.br/dataUse?lang=en for up-to-date information.

10. See http://splink.cria.org.br/indicators/index?setlang=en for more information.
11. See http://splink.cria.org.br/dc/index?&setlang=en for more information.
12. See http://inct.splink.org.br/index for more information.

References

Barjak, Franz, Kathryn Eccles, Eric T. Meyer, Simon Robinson, and Ralph Schroeder. 2010. "The Emerging Governance of E-infrastructure." Olten, Switzerland: Fachhochschule Nordwestschweiz. Hochschule für Wirtschaft.

Canhos, Dora Ann Lange, Mariane S. Sousa-Baena, Sidnei De Souza, Leonor Costa Maia, João R. Stehmann, Vanderlei P. Canhos, Renato De Giovanni, Maria Beatrix Machado Bonacelli, Wouter Los, and A. Townsend Peterson. 2015. "The Importance of Biodiversity E-infrastructures for Megadiverse Countries." *Community Page PLOS Biology.* http://dx.doi.org/10.1371/journal.pbio.1002204.

Canhos, Dora Ann Lange, Mariane S. Sousa-Baena, Sidnei Souza, Letícia Couto Garcia, Renato De Giovanni, Leonor Costa Maia, and Maria Beatriz Machado Bonacelli. 2014. "Lacunas: A Web Interface to Identify Plant Knowledge Gaps to Support Informed Decision Making." *Biodiversity and Conservation* 23 (1): 109–31. doi: 10.1007/s10531-013-0587.

Convention on Biological Diversity. 2010. *Strategic Plan for Biodiversity 2011–2020, Including Aichi Biodiversity Targets.* https://www.cbd.int/sp/.

David, Paul A. 2006. *Towards a Cyberinfrastructure for Enhanced Scientific Collaboration: Providing its Soft Foundations May Be the Hardest Part.* Cambridge, MA: MIT Press.

Gibbons, Michael, Camille Limoges, Helga Nowotny, Simon Schwartzman, Peter Scott, and Martin Trow, eds. 1994. *The New Production of Knowledge: The Dynamics of Science and Research in Contemporary Societies.* London: SAGE Publications.

Hobsbawm, Eric J. 1994. *Era dos extremos: o breve século XX, 1914–1991.* 2nd ed. São Paulo: Companhia das Letras.

Nowotny, Helga, Peter Scott, Michael T. Gibbons, and Peter B. Scott. 2001. *Re-Thinking Science: Knowledge and the Public in an Age of Uncertainty.* Cambridge, UK: Polity Press.

Sachs, Jeffrey D. 2015. *The Age of Sustainable Development.* New York: Columbia University Press.

World Commission on Environment and Development (WCED). 1987. *Our Common Future, From One Earth to One World.* United Nations (Report). http://www.un-documents.net/ocf-ov.htm#1.2.

United Nations Environment Programme. 1992. *Convention on Biological Diversity.* United Nations. https://www.cbd.int/convention/articles/default.shtml?a=cbd-08.

CHAPTER 7

Collaborative Development of an Open Knowledge Broker for Disaster Recovery Planning

Maurice McNaughton and Lila Rao-Graham

Abstract

Disaster Recovery Plans (DRPs) are costly but necessary for Small Island Developing States (SIDS) that are frequently affected by hurricanes and earthquakes. Using an approach based on Design Science, this project has sought to develop an Open Source Artifact that will streamline disjointed vocabulary and processes for disaster management between countries and across diverse stakeholders in the region. While revealing the complexities of creating open and enabling infrastructures, this project highlights that the social dimensions of building such tools are key to their long-term success. In that way, the success of "open" infrastructure should not be based on their design, but on the longer-term outcomes that they facilitate.

Introduction

In Jamaica and the Caribbean region, data produced using public resources are generally considered the private property of the agency that generated them, perhaps due to the perceived power of information conferred on the custodians. Cultural and institutional habits often forego the active sharing and use of data, and other forms of evidence, for policy and decision making. Aside from these cultural tendencies, there are also structural/institutional barriers arising from the limited scalability and resource endowments of the public administrations of Small Island Developing States (SIDS) that inhibit effective data sharing and use.

This purpose of this project was to determine whether a collaborative and shared approach could provide a solution and meet the need for a cost-effective and efficient Disaster Recovery Plan (DRP) in SIDS in the Caribbean. To this end, the project planned to develop an overall architecture (which we have termed a "knowledge broker") and the development of a shared vocabulary as a key sub-component of this architecture. This project is situated within an interesting domain for examining the characteristics, governance, and patterns of interactions within an open and collaborative environment. Given the Caribbean's vulnerability to, and experience with, natural disasters, there is a shared interest and strong regional commitment to collaboration around comprehensive disaster management and the sharing of knowledge resources, artifacts, and response coordination.

However, the research highlights that even with such a naturally conducive context toward open and collaborative knowledge solutions to common problems, there are other barriers that can limit the effectiveness of these open approaches. While there seems to be an active willingness to share knowledge resources, the primary challenges with the efficacy of this de facto "knowledge commons" are standardization, coordinated production, and having a good sense what knowledge resources already exist ("How do we know what we know?"). There is no central knowledge authority or directory that someone can go to in order to find out what resources are available. Thus, they continue to exist in silos with limited sharing.

In this chapter, we describe the development of a knowledge broker for the disaster management domain, an important component of which is a common, online, and interactive vocabulary. The development of this knowledge broker and by extension a common vocabulary requires the active engagement, participation, and ownership by the DRP community and is an iterative and progressive process. We discuss factors that influence the choice of the appropriate representation of the semantic concepts within a specific knowledge domain, as well as technology platform options. Ultimately, openness should not be regarded as an inherently advantageous "state," but rather the outcomes of openness within a particular knowledge context are what should be examined to determine its merits, specifically its influence on one or more of the individual freedoms associated with the philosophical notion of "Development as Freedom" (Sen 2001).

Background

Jamaica and the Caribbean region have enjoyed a long history of cultural, trade, and commercial "openness" as they benefit from a geographically advantageous location astride the major East-West shipping lanes and are blessed with deep, large natural harbours with short channels to open seas. Indeed, from the fifteenth century onward, the strategic importance of the Caribbean as a shipping and trading gateway between East and West has opened up the region to a multiplicity of social and cultural influences, adaptation, and assimilation from many different sources—European, African, Asian, and North American—resulting in a socio-cultural melting pot aptly exemplified by Jamaica's national motto "Out of Many, One People."

With several countries gaining independence since the 1960s, the post-colonial Caribbean generally enjoys a strong democratic tradition, constitutional and practised freedom of expression, including a very liberal (not in a political sense) and unimpeded press. Decades of north-bound migratory patterns, with persistent strong social ties to the home country and its cultural norms, together with its immense popularity as a tourist destination, mean that Caribbean culture and its icons (e.g., Bob Marley and reggae music) have had a disproportionate influence on global popular culture.

Does this rich, distinctively multicultural heritage make for what one might call an open society? Therein lies the paradox. The political leadership of the post-colonial, independent Caribbean has largely managed to spread a combination of externally imposed and self-inflicted layers of political and administrative bureaucracy across public administration. Professor Edwin Jones, widely recognized as the doyen of public-sector management in Caribbean societies, expresses this best as "institutional capture," imposed on the public-sector bureaucracy by Indigenous political actors and manifesting in a "happy cohabitation between politics and bureaucratic corruption which naturally leads to mal-administration" (Jones 2015). It is against this distinctively paradoxical Caribbean background that we interpret "openness" using elements of the Knowledge Commons framework[1] (Frischman, Madison, and Strandburg 2014).

Culture of Openness in Public Administration and Governance

In addition to the adopted tendencies that stand in the way of the active sharing described at the outset, there are also structural/institutional barriers arising from the limited scalability and resource endowments of the public administrations of SIDS that inhibit effective data sharing and use. Capacity-building efforts are required to effectively use new technologies to investigate, analyze, communicate, and inform policy and/or decision making.

Perhaps spoiled by a legacy of unbridled freedom of expression, the general civil society and the media seem apathetic with regard to demanding Open Data. Caribbean governments have certainly been slow to embrace formal Open Government/Open Data movements. Recently, however, there has been a more active thrust toward Open Data and Open Government initiatives with increased regional advocates supported by multilateral agencies.

Open Science is a broad concept that encompasses a multitude of assumptions about notions of knowledge creation and dissemination. Fecher and Friesike (2014) attempt to structure the overall discourse by proposing five schools of thought on Open Science:

1. the *infrastructure school* (which is concerned with the technological architecture)
2. the *public school* (which is concerned with the accessibility of knowledge creation)
3. the *measurement school* (which is concerned with alternative impact measurement)
4. the *democratic school* (which is concerned with access to knowledge)
5. the *pragmatic school* (which is concerned with collaborative research)

This chapter is most concerned with the *infrastructure school* and the way technological architecture fosters interaction among physically dispersed individuals and enables collaborative practices and knowledge sharing. Essential core capabilities of such an enabling infrastructure include: (1) management and sharing of knowledge objects for use and re-use; (2) incentives for knowledge producers to make their objects available; (3) an open and extensible environment;

and, (4) knowledge collaboration and sharing that is geared toward "action" rather than simply "storage and exchange" (see De Roure and Goble 2009). These attributes are readily applicable to the public administration and governance setting, and more specifically within the disaster management domain.

Context—Caribbean Disaster Management

The specific context within which our project is situated provides an interesting domain for examining the characteristics, governance, and patterns of interactions within a knowledge commons.

Currently, a number of institutions/entities in the region are developing documents and databases related to disaster management and recovery. Additionally, there are a number of experts in the area. However, although there is an active willingness to share these resources, there is no central knowledge authority or directory that someone can go to in order to find out what resources are available. Thus, they continue to exist in silos with limited sharing.

Community Members/Governance

The first step in developing this knowledge broker was to recognize that the success of the system was heavily reliant on the close collaboration of all the region's stakeholders. The Caribbean Disaster Management community is well organized, with the Caribbean Disaster Emergency Management Agency (CDEMA)[2] designated as a regional inter-governmental agency for disaster management in the Caribbean Community (CARICOM). CDEMA's mandate is to fully take up its role as facilitator, driver, coordinator, and motivating force for the promotion and engineering of Comprehensive Disaster Management (CDM) in all eighteen participating states. CDEMA is supported by, and actively engages with, a network of national disaster management agencies. For example, in Jamaica, the Office of Disaster Preparedness and Emergency Management (ODPEM)[3] is the main body responsible for coordinating the management of the various types of disasters, while in St. Vincent and the Grenadines, it is the National Emergency Management Office (NEMO)[4] that is assigned the role of activating the community on a countrywide basis to deal with disasters.

To return for a moment to Jamaica and ODPEM, the role of Regional Coordinators (RCs) is described as follows:

The RCs are primarily responsible for providing Parish Disaster committees, government agencies, private-sector organizations, and voluntary organizations with the necessary technical advice and assistance in implementing disaster preparedness measures.

Direct areas of assistance are:

- the development of parish and community disaster plans;
- the development and management of community disaster management committees, referred to as Zonal Committees; and
- the implementation of training programs such as shelter and emergency management (OPDEM).

CDEMA, therefore, represents a key knowledge actor within the commons and an important partnership over the course of the project. The formal community governance structures explicitly recognize the importance of gender and youth as active participants in CDM. This is an essential role given that DRP requires the collaboration of a number of stakeholders, including utility companies, government agencies, non-governmental agencies (NGOs), and the community. Other stakeholders include research groups, such as the Enhancing Knowledge and Application of Comprehensive Disaster Management (EKACDM) group,[5] which is working on a research project to implement the CARICOM Enhanced Comprehensive Disaster Management Framework, having as one of its key outcomes the creation of a regional network that generates, manages, and disseminates knowledge on disaster management.

Vocabularies

There is a great deal of value in representing the concepts of a domain as proposed in this research. In terms of the most suitable representation for these concepts, there are a number of options, including a glossary, a taxonomy, a thesaurus, or an ontology. Sometimes the distinctions between these mechanisms are not clear (van Rees 2003).

An ontology provides a formal description of a domain that can be shared among different applications and expressed in a language that can be used for reasoning (Gruber 1995; Noy 2004). It can also provide a framework for facilitating effective and efficient knowledge sharing by formally modelling the domain of discourse. Ontologies are typically viewed as presenting a shared understanding of some domain of interest, which is often conceived as a set of classes (concepts), relations, functions, axioms, and instances (Noy and McGuinness 2001). Noy and McGuinness (2001) highlight several benefits of developing an ontology to make domain assumptions explicit: (1) facilitating the sharing of a common understanding of the structure of information among stakeholders in a domain; (2) facilitating more effective communication and idea-sharing; (3) assisting new entrants in a field to quickly assimilate important domain concepts and knowledge; and (4) supporting the analysis of domain knowledge.

The thesaurus, on the other hand, includes a glossary of the terms, the definition of these terms, a hierarchical structure of the terms, and a link between these terms (e.g., synonyms). All of these options were considered for the DRP, and it was recognized that a more limited form of controlled vocabulary, a thesaurus, that is implementable using the simpler, but functionally competent, Simple Knowledge Organization System (SKOS) standards designed for structured, controlled vocabulary, was the most suited.

Knowledge Broker for DRP

This work focuses on the development of a knowledge broker for the DRP domain, an important component of which is a common, online, and interactive vocabulary. As such, it provides a technical solution for the integration of silos of knowledge related to DRP, which are currently dispersed throughout the region. This knowledge broker approach will provide a common semantic reference for resources distributed throughout the region and will facilitate a shared, collaborative approach to addressing DRP in the region. The first step in developing this broker was to recognize that the success of the system would be heavily reliant on the close collaboration of all of the regional stakeholders. Given that the Caribbean Disaster Management community is well organized and led by CDEMA, they were seen as the critical entity through which commitment to this project could be obtained. The objectives of this project were well aligned with those of CDEMA, which made it quite easy to form an alliance.

This broker framework provides an end-user with a single point of reference to search for DRP resources. These resources do not need to be physically integrated into one central repository but may reside where they were created and tagged with appropriate terms that describe what they represent in the DRP domain. Some queries that users can submit through the knowledge broker are illustrated in Figure 7.1. For example, if the end-user is interested in finding all resources related to hurricanes, that single query will be sent to the broker, which will then identify matching resources (i.e., documents, databases, entities, or experts). Additionally, if a new DRP resource becomes available, then it is important to tag this resource with appropriate DRP terms. Through the common, open, online, and interactive vocabulary, the knowledge broker will match the terms in the document with those of the vocabulary and identify the terms used to tag this resource.

Figure 7.1. The Knowledge Broker Framework

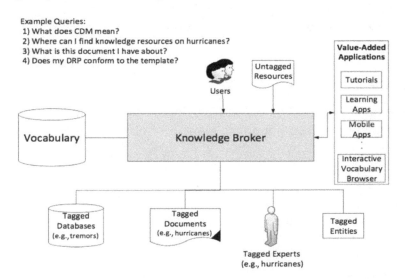

Given that an open approach and open technologies can be used in the development of this Knowledge Broker framework, it provides a tremendous opportunity for the development of value-added applications. An interactive vocabulary browser is being developed that will allow end-users, for example, to traverse and query the vocabulary and find the definition of terms and the synonyms for terms, given that we know different regions use different terms for the same thing. This vocabulary will become an Open Data source that will be available for developers so that other learning and mobile applications can be created that add value for the domain. This is an exciting opportunity for a region that is looking to provide development opportunities and solve regional problems, one of which is DRP.

The Design Science approach (Hevner et al. 2004; March and Smith 1995; Peffers and Gengler 2003) was used for the development of the knowledge broker for the DRP domain. The Design Science paradigm endeavours to extend the boundaries of knowledge by creating new and innovative artifacts (in this case the knowledge broker and DRP vocabulary) while seeking to improve and understand the behaviour of aspects of information systems through the analysis of the use and/or performance of such artifacts (Hevner et al. 2004). Hevner and others (2004) specify seven guidelines for conducting and evaluating good Design Science research; these guidelines were adopted for the research (see Table 7.1).

Table 7.1. Design Science Research Guidelines

Guideline	Description	Relevant Project Activities
Guideline 1: Design as an Artifact	Design Science research must produce a viable artifact in the form of a construct, a model, a method, or an instantiation.	The Knowledge Broker with a DRP vocabulary as a component is the artifact that will be produced from this research.
Guideline 2: Problem Relevance	The objective of Design Science research is to develop technology-based solutions to important relevant *issues*, including those of regional and national importance.	DRP is essential to Caribbean islands given they are prone to natural disasters. Additionally, given their limited resources and the fact they are susceptible to the same disasters, a collective approach to such an important regional problem is essential.

Table 7.1. Design Science Research Guidelines *(continued)*

Guideline	Description	Relevant Project Activities
Guideline 3: Design Evaluation	The utility, quality, and efficacy of a design artifact must be rigorously demonstrated via well-executed evaluation methods.	Various processes and mechanisms for evaluating the artifacts are proposed. This artifact will be evaluated using the Observational Method of a case study.
Guideline 4: Research Contribution	Effective Design Science must provide clear and verifiable contributions in the areas of the design artifact, design foundations, and/or design methodologies.	This research extends the existing body of research as it relates to Open Source solutions, ontologies, and DRP. It fills a void in the literature as it pertains to the need for more tools and technologies for the DRP domain.
Guideline 5: Research Rigor	Design Science research relies upon the application of rigorous methods in both the construction and evaluation of the design artifact.	This approach has been developed through building on related research and filling identified gaps.
Guideline 6: Design as a Search Process	The search for an effective artifact requires utilizing available means to reach desired ends while satisfying laws in the problem environment.	The justification of using an ontology for DRP has been clearly articulated and how this would lead to more effective DRP illustrated.
Guideline 7: Communication of Research	Design Science research must be presented effectively both to the technology-oriented as well as management-oriented audiences.	This research is being disseminated to both practitioners and academics. The key stakeholders have been engaged in the development process, thus ensuring it will be developed based on their needs. This work will be presented at academic conferences and documented through journal publications.

Source: Hevner et al. 2004

The development of the knowledge broker takes two converging paths. One path addresses the development of the DRP vocabulary, and the second path addresses the more technical issue of the design of the architecture being used for the implementation of the knowledge broker.

DRP Vocabulary

The DRP domain requires the collaboration of a number of stakeholders (e.g., utility companies, government agencies, NGOs). Additionally, in the Caribbean region, many countries face similar DRP issues and have limited resources to address them. This diversity of stakeholders and limited resources mean that there are tremendous opportunities if collaboration and sharing are used to tackle common regional problems. However, for this collaboration to take place effectively, all stakeholders must be using a consistent vocabulary. Implementing this common vocabulary as an online, open, and interactive shared resource can lead to a number of benefits, including the following:

- Facilitating the sharing of a common understanding in DPR, thereby reducing the possibility of confusion and ambiguity that may arise when different groups of stakeholders come together to share resources and make decisions (Altay and Green 2006). Given that the Knowledge Broker facilitates the integration and sharing of resources, it is possible that there is semantic ambiguity in the data, which can be addressed through the use of the vocabulary.
- Allowing for the automated evaluation of the DRP (e.g., checking for consistency of the plan).
- The vocabulary can also be used by countries or sectors wanting to develop DRPs to understand the important domain knowledge.

This development of a common vocabulary required the active engagement, participation, and ownership by the DRP community; the process itself is iterative and progressive and comprised the following steps:

1. Identification of the DRP knowledge resources
   a. identification and engagement of key stakeholders in the region of interest

      b. identification of key DRP resources (e.g., documents, data-bases, entities, and human expertise)

      c. identification of international standards and documents that address disaster management

2. The development of the initial vocabulary through the extraction of information from the stakeholders and documents

3. Feedback from stakeholders to update the vocabulary

4. Evaluation of vocabulary

The first stage of the development of this vocabulary was the identification and engagement of the key stakeholders in the region who have responsibility at various levels for disaster recovery planning. These include stakeholders at both the local and regional level and involve agencies established by the government as well as research initiatives and projects whose focus is on disaster preparedness and management in the Caribbean.

Once these groups were identified, they were engaged and able to describe the various knowledge initiatives within the region, sharing the existing documents and data sources that were useful in describing the domain. These documents included disaster plan templates, sector evacuation plans, and national plan models. These stakeholder engagements and the documents shared were used to extrapolate the various concepts and terms used in the domain and to understand the relationships between these concepts. They were also useful in gaining an understanding of the existing DRP practices (if any) and the concepts, terms, and activities currently being used locally, which is important to ensure regional consistency.

In parallel, it is important to identify and incorporate emerging international standards and benchmarks in the general disaster management domain. This will help to ensure the consistency and conformance of the regions' DRP practices, where applicable, with international best practice.

This process of the identification and engagement of stakeholders and resources was very interesting as it was extremely exploratory in nature. The starting point was connecting with an entity at the university where there was already an existing relationship. By engaging this entity, other entities and experts were identified who in turn identified still more. In engaging the stakeholders, care was taken to ensure that the objectives of this project could be aligned with the objectives of the organization or entity engaged. Once this alignment was made clear, there was a great deal of interest in collaboration. This made it quite easy to obtain the commitment of the stakeholders.

Once this understanding of the domain was completed, the vocabulary was developed and shared with a few of the stakeholders to ensure that it represented their needs. The vocabulary was then updated to reflect these needs. A portion of the vocabulary that was developed is shown in Figure 7.2. For each of the terms in the vocabulary, the term, synonyms for the term, and a definition of the term is represented. The synonyms are important for representing, among other things, different terms that are used to refer to the same thing depending on the region. For example, hurricanes, cyclones, and typhoons are different names according to the region they hit, but they all refer to the same phenomenon. The vocabulary was implemented based on emerging, linked Open Data standards to allow for seamless integration with other online semantic references (e.g., Climate Tagger[6]).

Figure 7.2. A Portion of the DRP Vocabulary

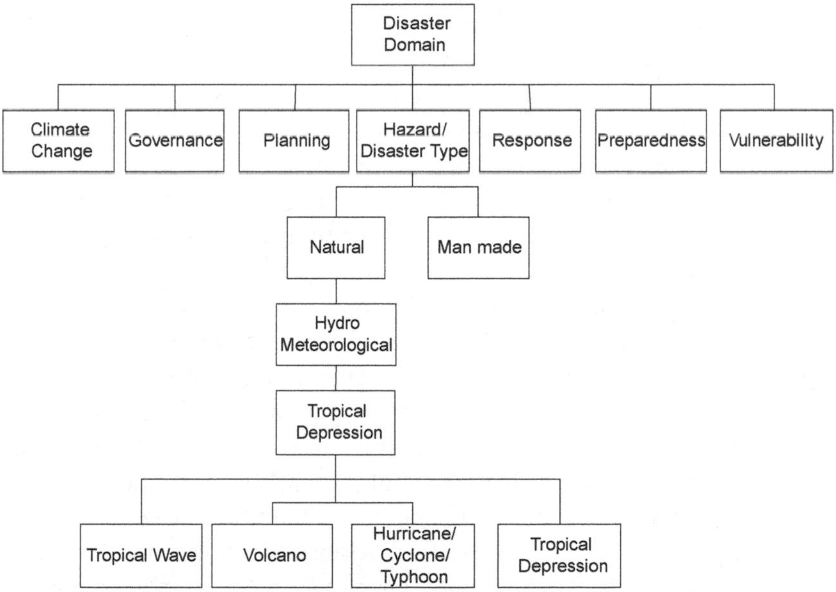

Source: CDEM, A National Emergency Management Plan.

The vocabulary is then made available to a larger group of stakeholders in an online, interactive way, allowing the stakeholders to query, traverse, and annotate the vocabulary to reflect any changes that they consider important. Finally, it will be made available to the community as an Open Source tool.

Knowledge Broker Architecture

A number of Semantic Web Platforms were examined to identify the technologies most suited for the application needs. The following functions were essential to the success of the DRP broker:

1. The implementation of the domain vocabulary. This vocabulary will represent the valid terms of the domain; thus, all the tags for the resources will be taken from this vocabulary.
2. The ability to extract the key terms from a resource by matching it against the terms of the vocabulary. This extraction process also facilitates the checking of the structure of a document (e.g., DRP plan) against what has been obtained and defined in the vocabulary.
3. The ability to automate the process of tagging the new resources that are to be added to the system. Not all of the resources may be automatically tagged and, therefore, some human intervention will be required. For example, the documents of the system can be tagged using an extraction process, but this will not be as easy for human experts. These resources will have to be tagged based on their areas of expertise as described in their biographies or by interviewing them.

Given the importance of collaboration and sharing of these resources as well as the need to offer these solutions in a cost-effective way, an open approach was used in the development of the platform and vocabulary. The shared vocabulary was defined as a semantic web thesaurus composed of concepts of the domain, which were expressed as triples using the Resource Data Format (RDF). Metadata and relations between concepts are defined in RDF using the SKOS. Two important components of the system were the Thesaurus Manager and the Thesaurus Explorer. The Thesaurus Manager allows the domain expert to create and manage the thesaurus's concepts without editing RDF files directly. The Thesaurus Manager used in this architecture is VocBench 2.[7] VocBench 2 is a free and Open Source web-based thesaurus manager that runs on an Apache Tomcat server and uses MySQL as a database management system and Semantic Turkey[8] as an RDF triple store. The

Thesaurus Explorer allows a user to view and comment on the concepts in the thesaurus. It is a graphical user interface that can display the thesaurus in different ways (e.g., a hierarchical tree containing narrower and broader concepts). The Thesaurus Explorer in this architecture builds on the SKOS Play Open Source platform, which is a Java application deployed on Apache Tomcat. It uses a SPARQL endpoint to query the thesaurus and display the results graphically. A SPARQL endpoint enables users (human or other) to query a knowledge base via the SPARQL language.

The three-layer architecture used is outlined in Figure 7.3. The first layer, VocBench, will be used as the Thesaurus Manager for the vocabulary. It comes with a user interface that enables the user to upload their thesaurus as an RDF file. From the interface, VocBench provides the thesaurus, which can be queried and edited. The administrator can update the publicly visible data VocBench and also export the data as various formats. The choice will be skos-xl. Exporting it in this format produces a file that has been formatted so that it is compatible with applications that use SKOS data models, namely SKOS Play. The second layer takes a formatted RDF file and a running Fuseki[9] server. The interface provided by Fuseki enables users to upload datasets, query them, and also expose the dataset through a SPARQL endpoint. Once a dataset is uploaded successfully, the SPARQL endpoint is readily available and exposed. The third layer assumes that the SPARQL endpoint provided by Fuseki can now be used with the front-end application, SKOS Play.

Figure 7.3. Knowledge Broker Architecture

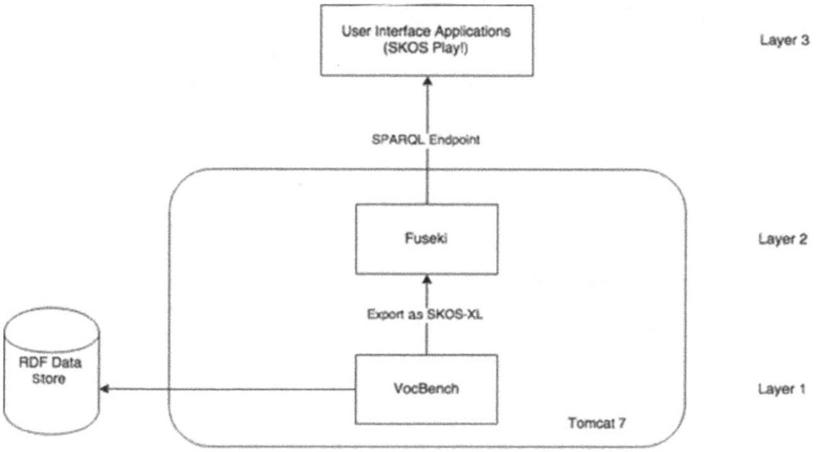

These technologies will be integrated to develop the knowledge broker for DRP.

Lessons Learned on Openness

Deliberations on the nature of "openness" by the Open Scholarship Initiative[10] have identified several attributes that may be used to characterize the degree of openness of any specific resource (Anderson et al. 2016). The resulting DART framework is based on the premise that "open" is not a binary state. Rather, it is a spectrum that exists on multiple dimensions—specifically: Discoverability, Accessibility, Reusability, and Transparency. A summary of these attributes is provided in Table 7.2.

Table 7.2. DART—Dimensions of Openness

Dimension	Attributes	Comments
Discoverable	Extent to which a resource: – is indexed by search engines; – has sufficient, good quality discovery metadata; – has persistent unique identifiers and – has explicit rights statements	This may be the most fundamental baseline condition of openness: if an object is not discoverable, then it cannot be considered open.
Accessible	Free (in terms of cost) to all users at point of use, in perpetuity; downloadable; machine-readable timeliness of availability	These are the attributes most commonly associated with open resources (software, data, etc.),[11] although variations exist based on various licensing conditions
Reusable	Usable and re-usable (including commercial uses); modifiable and able to be further disseminated	
Transparent	Peer review; impact metrics, transparency in the research process; author transparency (funding source, affiliations, roles, and other disclosures such as conflict of interest)	Provides the potential user of the works a means of assuring quality, integrity, and source

While this DART framework was derived primarily in the realm of Open Access as it relates to scholarly publishing, it provides a useful and convenient lens to summarize our insights from this initiative.

Discoverability: This has been, perhaps, the most evident gap in relation to disaster management knowledge resources in the Caribbean (how do we know what we know?). Although the Caribbean Disaster Management community has well-organized regional governance mechanisms and a visible, active commitment to collaboration and sharing of knowledge, the absence of a central knowledge authority or directory has severely limited the discoverability of knowledge resources. The Knowledge Broker developed through this initiative has been demonstrated to be an effective mechanism for the integration of DRP knowledge silos currently dispersed throughout the region. The open, semantic functionality of the Knowledge Broker also provides a platform with standardized APIs (SPARQL endpoint) that will encourage the development of additional discovery aids such as search engines and metadata directories.

Accessibility: In addition to being limited by discoverability, many of the existing knowledge assets exist in the form of off-line documents that rely on a knowledge of, and relationships with, existing custodians to secure access to the documents. The design of the Knowledge Broker not only allows for indexed reference to online knowledge sources, but also provides its own content pool, allowing existing off-line documents to be uploaded, tagged, and indexed, thus enhancing accessibility.

Reusability: It is noteworthy that efforts were made by CARDIN[12] as early as 2000 to develop a Controlled Vocabulary on Disaster Information (Lashley, Henry, and Caribbean Disaster Information Network 2000). The resulting artifact in the form of a PDF document had limited re-use and utility and, indeed, became a non-discoverable knowledge resource itself as institutional memory of its existence faded. The semantic web technologies and open standards used in the development of this Knowledge Broker significantly enhance the potential for re-usability of existing CDM knowledge assets through developers creating a variety of learning and mobile applications that ultimately amplify the value of the domain knowledge.

Example applications produced as part of the initiative include the previously mentioned interactive vocabulary browser that allows end-users to traverse and query the vocabulary, which itself represents an Open Knowledge resource that is available for re-use.

Transparency: Transparency and quality assurance in this context began with an extensive dialogue and consultation with the Caribbean DRP community to ensure active engagement, involvement, validation, and ownership of the controlled vocabulary that underpins the knowledge broker. On an ongoing basis, the submission and tagging of documents by the individual members of the disaster community under CDEMA's overall stewardship provide a degree of peer-review support under benevolent governance that approximates the quality assurance process employed so effectively in the Open Source software domain.

Conclusion

The knowledge broker provides a technical solution for integrating silos of DRP knowledge that are currently dispersed throughout the region. This Knowledge Broker approach provides a common semantic reference for resources and will facilitate a shared, collaborative approach to addressing disaster recovery planning in the region. However, the technical artifact alone should not be viewed as a panacea; it has to be coupled with the right social dynamics in order to build sustainable knowledge communities.

Ultimately, openness should not be regarded as an inherently advantageous "state," but rather the outcomes of openness within a particular knowledge context are what should be examined to determine its merits. For instance, well-known open paradigms generally make reference to the "freedoms" that arise as a result of openness; Open Source software grants the user of the software access to source code and four freedoms: to use, copy, study and modify, and improve and redistribute the software. Perhaps we might eventually consider assessing the openness of Caribbean societies through an examination of its influence on the individual freedoms associated with Amartya Sen's notion of "Development as Freedom," that is, political freedoms, economic facilities, social opportunities, transparency guarantees, and protective security (Sen 2001). This is an intriguing idea that warrants further discourse. In at least two respects, the Senian viewpoint emphasizes the multidimensionality of development and also debunks the notion of development as a supply-side phenomenon. This resonates well with our concluding remarks on "openness" as being non-binary and contingent rather than normative. We expect that the effective assimilation and utilization of the open knowledge broker by the Caribbean disaster management community will significantly enhance the usability and utility of disaster management knowledge assets and ultimately impact positively on one or more of these development freedoms.

Notes

1. See http://knowledge-commons.net/publications/gkc/research-framework/.
2. See http://www.cdema.org/.
3. See http://www.odpem.org.jm/.
4. See http://www.security.gov.vc/index.php?option=com_content&view=article&id=12&Itemid=5.
5. See http://www.uwi.edu/EKACDM/index.aspx.
6. See http://www.climatetagger.net/.
7. See http://vocbench.uniroma2.it/.
8. See http://semanticturkey.uniroma2.it/.
9. See https://jena.apache.org/documentation/fuseki2/.
10. See http://journals.gmu.edu/osi/index.
11. See the Open Definition at http://opendefinition.org/od/2.1/en/.
12. Caribbean Disaster Information Network.

References

Altay, Nezih, and Walter G. Green. 2006. "OR/MS Research in Disaster Operations Management." *European Journal of Operational Research* 175 (1): 475–93.

Anderson, Rick, Seth Denbo, Diane J. Graves, Susan Haigh, Steven Hill, Martin Kalfatovic, Roy Kaufman, Catherine Murray-Rust, Kathleen Shearer, Dick Wilder, and Alicia Wise. 2016. "Report from the 'What Is Open?' Workgroup." *Open Scholarship Initiative Proceedings* 1: 1–5.

De Roure, David, and Carole Goble. 2009. "Software Design for Empowering Scientists." *IEEE Software* 26 (1): 88–95.

Fecher, Benedikt, and Sascha Friesike. 2014. "Open Science: One Term, Five Schools of Thought." In *Opening Science*, edited by Sönke Bartling and Sascha Friesike, 213–24. Cham, Switzerland: Springer International Publishing. https://doi.org/10.1007/978-3-319-00026-8.

Frischman, Brett M., Michael J. Madison, and Katherine J. Strandburg. 2014. *Governing Knowledge Commons*. New York: Oxford University Press.

Gruber, Thomas R. 1995. "Toward Principles for the Design of Ontologies Used for Knowledge Sharing." *International Journal of Human-Computer Studies* 43 (5–6): 907–28.

Hevner, Alan R., Salvatore T. March, Jinsoo Park, and Sudha Ram. 2004. "Design Science in Information Systems Research." *MIS Quarterly* 28 (1): 75–105.

Jones, Edwin. 2015. *Contending with Administrivia: Competition for Space, Benefits and Power.* Kingston, Jamaica: Arawak Publications.

Lashley, Beverley, Houple Henry, and Caribbean Disaster Information Network. 2000. *Controlled Vocabulary on Disaster Information (Vocabulario Controlado Sobre Desastres).* Kingston, Jamaica: Caribbean Disaster Information Network (CARDIN).

March, Salvatore, and Gerald F. Smith. 1995. "Design and Natural Science Research on Information Technology." *Decision Support Systems* 15 (4): 251–66.

Noy, Natalya F. 2004. "Semantic Integration: A Survey of Ontology Based Approaches." *SIGMOD Record* 33 (4): 65–9.

Noy, Natalya F., and Deborah L. McGuinness. 2001. "Ontology Development 101: A Guide to Creating Your First Ontology." *Stanford Knowledge Systems Laboratory Technical Report KSL-01-05* and *Stanford Medical Informatics Technical Report SMI-2001-0880.*

OPDEM. Office of Disaster Preparedness and Emergency Management. n.d. www.odpem.org.jm/AboutUs/OrganizationStructure/tabid/90/Default .aspx

Peffers, Ken, and Charles E. Gengler. 2003. "How to Identify New High-Payoff Information Systems for the Organization." *Communications of the ACM* 46 (1): 83–8.

van Rees, Reinout. 2003. "Clarity in the Usage of the Terms Ontology, Taxonomy and Classification." *CIB REPORT* 284 (432): 1–8.

Sen, Amartya. 2001. *Development as Freedom.* New Delhi: Oxford University Press.

CHAPTER 8

Harmonization of Open Science and Commercialization in Research Partnerships in Kenya

Maurice Bolo, Victor Awino, and Dorine Odongo

Abstract

When public universities partner with commercial industries for re-search purposes, there is the potential for great synergy but also for ideological conflict. In recent years, Kenyan universities and research institutions have seen the simultaneous growth in both pro-Open Science policies, as well as an increased pursuit of knowledge pat-ents. This project sought to assess the national and institutional policy context for the potential of Open Science, and what this shift could entail for partnerships between public and private entities. Through an assessment of three case studies, the project concludes that while the country has strong policy guidance around the importance of Open Science and access, the nitty-gritty details of "who owns what" remain an obstacle for true collaboration between institutions and across sectors.

Introduction

Kenya's aspiration to transition to a knowledge-based, middle-income country is aptly captured in its long-term development blueprint — the Vision 2030. This Vision is hinged on science, technology, and innova-tion (STI) in the country's foundation for socio-economic development. This enhanced role for knowledge in economic development has thrust the institutions of higher learning and research into the centre of the country's development agenda, and placed renewed emphasis on the need for closer collaboration between the universities and private sector to enhance the commercialization of research findings. This call for closer collaborations and partnerships leads to a number of problems/concerns. First, it has opened up hitherto hidden cultural tensions between academic traditions with its emphasis on Open Science as a public good and a commercial culture that emphasizes privatization of knowledge. These contradicting approaches to Open Science and commercialization are likely to undermine the role that universities/Public Research Institutions (PRIs) play in undergirding Kenya's transition to a knowledge-based economy. Secondly, a lack of guiding policies and principles on how to harmonize the said cultural contradictions affects researchers' ability to disseminate and exploit their findings, build on their current work, conduct follow-up research and innovation, and participate in new collaborations.

Resolving these conflicts requires a broad institutional and governance framework that not only makes these potential conflict areas explicit, but also sets out principles and guidelines on how to minimize and manage such conflicts in a progressive manner. Multi-disciplinary research partnerships by definition bring together actors of diverse backgrounds in terms of disciplines, culture, ethics, and tradition. Ensuring that the aspirations of all these partners and their diverse practices operate in harmony requires intentional efforts at trust building. This calls for the need to manage the different cultural expectations of the various partners.

At the national policy level, the government has instituted a number of measures to support both the generation and sharing of knowledge from publicly funded research as well as commercial exploitation and private-sector uptake of the same. For example, to support research and innovation, the NACOSTI established the STI Competitive Grants Scheme in 2008 as a vehicle to fund multi-disciplinary and multi-institutional research partnerships. Since its inception, the government has progressively increased the research funding and broadened its thematic foci.

In 2013, the government repealed the *Science and Technology Act* (cap 250) and created three autonomous institutions to manage research and innovation. In this legislative shake up, the National Commission for Science, Technology and Innovation (NACOSTI)—formerly the National Council for Science and Technology (NCST)—was created to set the national research agenda and provide licensing and quality assurance functions. The National Research Fund (NRF) was charged with resource mobilization and funding. It now manages the STI Grants Scheme while the Kenya National Innovation Agency (KENIA) handles the promotion of research translation and has the responsibility of identifying, characterizing, and supporting Kenyan innovations. Additionally, the government has anchored the protection of intellectual property rights in the country's Constitution as well as enacting other enabling legislation to facilitate commercialization.

In response to the evolution at the national policy level, academic and research institutions are also setting up their own institutional policies. Nearly all public universities now have intellectual property rights (IP) policies and attendant IP management or technology transfer offices (TTOs). Most universities have also revised their governance structures to include positions of Deputy Vice Chancellors (DVCs) in charge of research and innovation. Interfacing with clients and communities has become a priority, and outreach has also become a key activity within the academic and research establishments. Many universities now have extension/liaison offices intended to be the "customer contact point" and manage their collaborations, especially with the private sector and other external partners.

Amid this evolving policy environment, key questions for this chapter remain: How has the potential cultural conflict manifested in research partnerships in Kenya? How have these conflicts affected the choices, practices, and behaviour of researchers involved in collaborative research projects? How have the national and institutional policies provided a mechanism for addressing the conflicts and where are the governance gaps? What measures should be undertaken to harmonize the Open Science policies with the need for commercial exploitation?

To answer these questions, the chapter draws on three case studies of government-funded contemporary research partnerships to discuss the challenges that researchers face. In summary, Case Study 1 highlights the initial development of a patented herbal food supplement by a private-sector company. The product required further validation, and a consortium comprising public universities, research institutes, and the private company was funded by NACOSTI to undertake the validation. However, the consortium failed to sign an agreement on IP rights, publication guidelines, and data protection and ownership. In the end, the private company applied and obtained IP rights over the new data and went ahead to develop and commercialize other products. In Case Study 2, a consortium of public universities and research institutes set out to research African indigenous vegetables for their commercial and climate resilience potential. No agreement on project management or partnership conflict management was signed, and there was no prior consideration as to the potential conflicts that might arise should any of these products prove to be commercially viable or lead to any novel findings. While at the time of this case study, the issue had not arisen, it was clear that the consortium was ill-prepared to deal with it. Finally, in Case Study 3, we present the development of a range of edible products used for health management and diversification of household income streams, which led to conflicts around budget and roles at the consortium level. As a result, a researcher from a public research institute is alleged to have used information and data from their consortium to negotiate with other partners and seek funding elsewhere, based on ideas that were initially developed collectively by the consortium.

The rest of the chapter proceeds as follows: From the introduction above, the chapter expounds on the overall methodology and criteria for selecting the case studies. The actual case studies and experiences are then presented. This is followed by exploring the national policy context in Kenya and discussing the governance framework for Open Science and intellectual property rights. It then proceeds to the institutional level and discusses policy and governance structures, and how these have affected the performance in patenting and publications in Open Access (OA) journals. These issues are then discussed in light of the policy and the institutional regime, the governance framework, and finally the performance and behaviour, as well as the choices and practices, of the researchers. The chapter ends with a short conclusion and recommendations.

Methodology

The study was conducted through the following approaches: (1) stakeholder interviews; (2) discussions with experts; and (3) case studies. These three approaches were complemented by a literature and documentary review, as well as an empirical desk research and institutional review of intellectual property rights and OA journal publications. In particular, we conducted an extensive literature and documentary review of the national and institutional policy environment to highlight the evolution of Open Science and commercialization approaches over time. We analyzed institutional accounts to examine governance and organizational changes in support of the transition to knowledge-based economies and their general responses to policy stimuli in the broader national, regional, and international contexts. We delved into institutional databases of the various academic institutions, especially their OA repositories, to analyze the development of OA policies and publications, and dug into the three-decades-long patent database of the Kenya Industrial Property Institute (KIPI) to draw out the trends on patent applications from public universities and research institutes.

After the literature review, we engaged in key informant interviews with selected stakeholders and experts on their experiences (with current practices and processes) and expectations as to how current challenges could be addressed. These interviews were executed through individual face-to-face interviews, as well as through a series of focus group discussions (mainly targeting researchers and research managers). Finally, we focused on three representative case studies to elicit the practical experiences of researchers involved in multi-disciplinary research partnerships. The primary respondents for these in-depth case studies were the principal investigators and co-principal investigators. Their responses were cross-checked by interviewing their partners and research/grant managers at their institutions.

Case Studies: Selection Criteria

Our initial idea was to use contemporary case studies derived from joint patent applications submitted to KIPI (1990–2013). This changed considerably once the study began, as a number of practical challenges emerged. We had assumed that joint applications submitted to KIPI would have resulted from collaborative research, and there would be sufficient background and contact details to select appropriate cases. We also assumed that participants would be willing to share their experiences. Our assumptions did not hold, as a considerable number of the joint applications involved international partners with limited contact details. As for local partners, most of the addresses and contact details in the KIPI registry did not lead us to the applicants. Similarly, given the sensitivity and secrecy surrounding intellectual property rights, some partners were unwilling to talk about their patents and inventions. In the end, we shelved the idea of using the joint patent applications from KIPI.

Another set of challenges was related to delays in securing cooperation from institutions managing research partnerships. Several attempts were made to secure the necessary authorization from both CNHR and LIWA (Linking Industry With Academia). However, at the time of this study, both institutions were undergoing leadership transitions and responses were quite slow. Due to the delays, we also dropped the case studies from CNHR (Consortium for National Health Research) and LIWA.

These challenges left us with the research consortia being funded under the NACOSTI STI grant schemes. Courtesy of an existing memorandum of understanding (MoU) between the Scinnovent Centre and NACOSTI, we obtained a database of their funded projects between 2008 and 2013. From this database, ten projects were initially selected based on the following criteria:

1. economic sectors—mainly agriculture, health and natural products, energy, and ICT;
2. the lead applicant—whether universities or public research institutes;
3. the likelihood of IP protection—whether patentable products/ processes were anticipated;
4. the status of the project—whether sufficient progress had been made to enable analysis; and
5. the extent and role of the private sector and/or other non-academic actors.

The research team perused the physical hard copy files of the ten projects to obtain:

1. the abstracts/summaries of the projects, including their objectives, outputs, and expected outcomes;
2. the consortia partners and their contact details;
3. the progress reports, what had been achieved, and the challenges;
4. proposed governance arrangements and role sharing; and
5. any considerations on intellectual property, benefit sharing, and publications, as well as data-sharing policies.

Case Studies and Practical Experiences

At the end of the study, six of these case studies had been conducted. However, only three are reported here due to their direct relevance to the theme of the current book. Details of the case studies are presented below.

Case Study 1: "From Sunguprot to Super Sunguprot": A Case of Follow-on Innovation

Sunguprot is a herbal food supplement with both anti-retroviral and nutritive properties. It comes in the form of porridge and is ideal for people suffering from HIV/AIDS, the malnourished, and the aged. *Sunguprot* was initially an invention of a private-sector company that had already obtained IP protection (utility model) and regulatory approvals from the Kenya Bureau of Standards (KEBS) and the Pharmacy and Poisons Board (PPB) to sell and market the products as food supplements. However, the products still required validation that necessitated further physio-chemical, micro-chemical, clinical, and pharmacological analyses to determine their safety, quality, and efficacy prior to producing prototypes and moving into large-scale commercialization.

A consortium consisting of a public university, two public research institutes, and the private-sector company sought to improve the functions and design of the production process of *sunguprot* and porridge as food supplements by obtaining data that aims not only at validating the products and processes, but also at developing agronomic strategies for sustainable production of *Tylosemafassoglensis*, an important ingredient in the products and one of the least studied plants. Despite several attempts, the partners failed to sign a consortium agreement that would provide guidance on IP rights, publication guidelines, and data protection issues. When the research was finalized and dissemination planned, the private-sector actor feared losing both the current data as well as his initial invention through public disclosure of the research findings. With no guidelines on how to resolve the IP rights issue and no agreement defining partners' obligations, the private-sector actor sought and obtained approval from the funder (NACOSTI) to apply for IP rights over the research findings. He applied and got protection over all the data from the research, and proceeded to develop "super *sunguprot*" as a superior product based on the research findings. Moreover, he developed other products based on the data, including an immune modulator (*canoma*) and *Sungu* lemonade herbal tea.

Case Study 2: "We Shall Wait and See": Research Consortia Develop Commercializable Products but Have No Idea Who Owns the Products

A multi-disciplinary consortium of two public universities, a public research institute, and an NGO sought to investigate the issues of climate resilience, sustainable production, value addition, and commercialization of African Indigenous vegetables. They focused on three species, namely: amaranth (locally known as *terere*); nightshade (locally known as *managu/osuga*), and the spider plant. In the consortium, the lead partner (a public university) was in charge of agronomy, product development, seed production, information dissemination, and socioeconomics while the collaborating university was in charge of physiology and climate modelling. The public research institute and the NGO were in charge of farmer mobilization and training. Even though these roles were stipulated in the proposal document, it ended at that point. No binding agreement was reached that would ensure that the partners delivered on these roles, or that, once funding was secured, no partner would shortchange the others. The institutional governance aspects of the project and any dispute resolution mechanisms were not factored in the consortium management structures. Further, the consortium hoped that in the end, they would have developed an "innovation centre of excellence" to bring together actors beyond the research fraternity to share knowledge, skills, and expertise. The information at the core of this centre of excellence would be gathered from farmers and other users of traditional knowledge with regard to recipes using various Indigenous vegetables and made available through Open Source by the centre.

The project has PhD and MSc students undertaking specific studies on food formulation and production, along with nutrient analysis, and coming up with new products, such as biscuits, doughnuts, and bread. While some of these are already being tasted and tested by selling to the students in the university canteens, the consortium has not defined ownership. When asked, they simply replied, "We shall wait and see if anyone claims ownership." Although by design the research was to lead to new products, neither the funders nor the researchers had prior consideration as to the potential conflicts that might arise should any of these products prove to be commercially viable or lead to any novel findings.

Regarding publications and OA, the consortium members noted that authorship would be based on (1) the lead partner for the particular specific objective and (2) the contribution of each researcher to the paper being published. However, there are no clear mechanisms to determine "extent of contribution." Interviews with the principal investigator (PI), for example, indicated that she prefers a model she experienced in a European Union-funded project whereby the youngest collaborator in the consortium becomes the first author and the most experienced becomes the last author. While this is intended to support and promote younger researchers, it may not be welcome in the local context, especially by the senior academics. Though at the time of this interview, the issue had not arisen, it was clear that the consortium was ill-prepared to deal with it.

Case Study 3: "They Ran With Our Knowledge": A Case of Post-Partnership Collaborations

In this project, *Manihotesculenta* (cassava), *Eleusinecoracana* (finger-millet), *Sesamumorientale L.* (simsim), *Chrotalariuochroleuca* (slender-leaf mild), *Chrotalariabrevidens* (slenderleaf bitter), and *Arachishypogaea* (groundnuts) were used to produce cookies, pre-cooked flour, noodles, crackers, and vegetable simsim products. Proximate composition, micronutrient, anti-nutrients, and food safety tests were done on the raw materials and on the final product prototypes produced on formulations based on nutritional values. The formulated product prototypes were then packaged, and their acceptability analyzed through organoleptic tests. Complete analyses of the products for nutritional and microbial levels have also been done. The project partners intended that the products be used for health management by the sick, elderly, children under five, and women of child-bearing age. Besides health management, these products are aimed at enhancing diet variety and diversity. Efforts are being made to introduce the products in the market through existing community groups, and to establish market structures for the products that will contribute to diversified income streams for households.

A disagreement arose in the research consortium concerning the sharing of resources (mainly budgets and roles). Given that these were not defined upfront, when the funding came, some of the partners felt they deserved more. In the absence of a conflict resolution mechanism or a binding consortium agreement, a researcher from a public research institute is alleged to have used information and data from their consortium to negotiate with other partners and seek funding elsewhere. In the absence of any guiding or binding contracts (or other instruments of governance), sharing ideas and knowledge freely exposes research partners to their knowledge being used without reference to them. When partners share ideas and knowledge in a project proposal, it is assumed that such knowledge is collectively owned. However, it is not clear what partners can and cannot do, as well as the timeframes within which any data/information/knowledge cannot be used without prior approval from consortium members.

Contrasting Policy and Practice: Contextualizing Case Studies Within the National Policy Context

The Kenyan STI policy framework anticipates that universities and public research institutes take the lead in the generation of technologies and inventions and transferring the same to the private sector and other beneficiaries. In "Sessional Paper No. 5 of 1982 on Science and Technology for Development," the government asserts that "the research in Kenya should lead to techno-economic feasibility and social acceptability of its innovations, construction of pilot plants and full-scale production." The government further undertook "to establish the linkages between universities and other institutions of higher education and the research establishments in government departments and industry." Similarly, in "Sessional Paper No. 1 of 1994," industry is encouraged "to develop mutually beneficial contractual links with the research institutes for the generation of viable technologies." These policies provide the general framework for multi-disciplinary research and academia–private-sector research collaborations. As noted also in "Sessional Paper No. 2 of 1996 on Industrial Transformation," there exists a weak linkage between the Kenyan industry and the research institutions, and "no structured mechanism exists for identifying problems of private industrial sector, which are then passed on to R&D institutions for investigation and formulation of appropriate solutions."

The STI policy and strategy (2008) sought to correct this situation and "encourage and support collaborative, multi-disciplinary scientific research in universities and other academic, scientific and engineering institutions." The STI policy advocates to "increase public investment for universities, government laboratories and research institutes to enable access to facilities and equipment needed for research for focusing on identified national strategic priority areas." In guiding its transition to a knowledge-based economy, "Sessional Paper No. 10 of 2012 on Kenya Vision 2030" gives primacy to the role of science, technology, and innovation and notes that Kenya Vision 2030 (2007) "recognizes the role of science, technology and innovation (STI) in a modern economy, in which new knowledge plays a central role in boosting wealth creation, social welfare and international competitiveness." The Sessional Paper further identifies four elements that allow effective exploitation of knowledge as:

1. an economic and institutional regime that provides incentives for the efficient use of the existing knowledge, the creation of new knowledge, and the flourishing of entrepreneurship;
2. an educated and skilled population that can create, share, and use knowledge well;
3. a dynamic information and communication infrastructure that can facilitate processing, communication, and dissemination; and, finally,
4. an effective innovation system (i.e., a network of research centres, universities, think tanks, private enterprises, and community groups) that can tap into the growing stock of global knowledge, assimilate and adapt it to local needs while creating new knowledge and technologies as appropriate.

National Policies and Open Science Governance

Interviews with researchers, policy makers, and other key stakeholders reveal that in the Kenyan context, "openness," "open science," and "open access" are considered with regard to the extent to which involved actors can: (1) access and share research facilities and infrastructure; (2) share information in designing and executing projects within teams/consortia; (3) disseminate information through publications and other events, whether jointly or individually; (4) freely participate in research collaborations with other parties beyond current partners; and (5) share benefits from commercializable research outputs. In order to examine how these Open Science aspirations are manifested in practice, we analyzed the three case studies against stated policies, governance arrangements, and performance at three levels: the national level, organizational level, and partnership/consortium level.

At the national level, our analysis showed that the need for openness and unrestricted access to information and knowledge is anchored in Kenyan laws, policies, and economic development blueprints. The country's supreme law—the Constitution of Kenya (2010)—recognizes the role of science and technology in its development endeavours and provides in article 11 that the State shall: (1) recognize the role of science and Indigenous technologies in the development of the nation; and (2) promote the intellectual property (IP) rights of the people of Kenya. Specific to openness, the Constitution provides for both freedom of expression and access to information in its bill of rights. Article 33 provides for freedom of expression including "academic freedom and freedom of scientific research." Similarly, article 35 deals with access to information and provides that "every citizen has the right of access to (a) information held by the state; (b) information held by another person and required, or the exercise or protection of any right or fundamental freedom...; and (c) that the state shall publish and publicize any important information affecting the nation." The Constitution therefore provides a broad framework within which to situate open and collaborative projects. The provisions for academic and scientific freedom, access to information, and the requirement for the State to make open and publish information are particularly relevant for increased openness in research collaborations. In this regard, it is important to recognize that the government has undertaken steps to realize the openness envisaged in its policies, and, in 2011, initiated the Kenya Open Data Initiative (KODI) under the Ministry of Information with the key objective of making government documents, databases, policies, and programs readily available to the public.[1]

Kenya operates a multi-agency regulatory framework with governance of IP and Open Science spread across different ministries and regulatory agencies. For example, in IP protection, KIPI, which is responsible for patents and in charge of implementing the *Industrial Property Act*, 2001, is domiciled at the Ministry of Industrialization and Enterprise Development. Copyrights are handled by the Kenya Copyrights Board (KECOBO) through the *Copyrights Act*, 2001 and housed at the Office of the Attorney General. Issues of plant varieties are handled by the Kenya Plant Health Inspectorate Service (KEPHIS) under the Ministry of Agriculture, while material transfer agreements (MTAs), which might be needed for research purposes, are handled by the National Environment Management Authority (NEMA) under the Ministry of Environment and Natural Resource Management. Issues of research, quality, and collaborations fall under the merit of the Ministry of Education, Science and Technology with two key regulatory institutions: National Commission for Science, Technology and Innovation (NACOSTI), and the Commission for University Education (CUE).

Institutional Policies and Governance for Open Science

At the institutional level, nearly all the universities have revised their charters to include the office of DVCs in charge of research and development (R&D). These offices are largely responsible for research strategy, quality, and outreach. Part of their roles includes linkages with external partners such as community and industry, and promoting collaborative research, contract research, and consultancies. Following the adoption of institutional IP policies, universities have created offices to manage the intellectual assets emanating from staff and students. These include the Intellectual Property Management Offices (IPMOs) and/or the Technology Transfer Offices (TTOs). These offices are supported by IP management committees or IP management boards whose roles include identifying potential novel products from research and organizing them for disclosure and protection. Regarding OA, most of the universities have established OA policies and digital repositories as well as embraced OA publishing since 2012. In most cases, the librarians are in charge of ensuring implementation and adherence to OA policies and principles. Figure 8.1 and Table 8.1 present the trends in adoption of institutional OA and IP rights policies as well as institutional repositories.

Figure 8.1. Establishment of IP and Open Access Policies at Kenyan Universities (2004–2015)

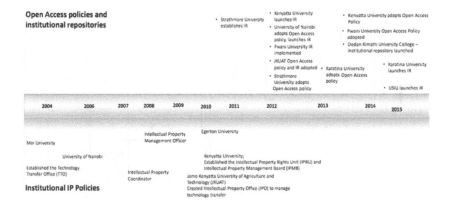

Table 8.1 History of Open Access and Intellectual Property

Related Policies at Kenyan Universities

University	Open Access Policy/ Repositories	IP Related Policy
University of Nairobi	Open Access Policy[2] adopted in 2012 and a digital repository[3] established in the same year.	It has a Research Policy[4] and has created the Office of the Deputy Vice Chancellor (Research and Development). To enhance commercialization, the Research Policy emphasizes the need to link research to commercialization by establishing Science and Technology Parks (STPs). The university established its Intellectual Property Policy[5] in 2006 and created the Intellectual Property Management Office (IPMO).
Kenyatta University	Open Access Institutional Repository Policy[6] adopted in 2014. Also has a digital repository and its content freely accessible via the Repository's web site.[7]	Intellectual Property Policy[8] and also created the Intellectual Property Rights Unit (IPRU) headed by a Director and assisted by the Intellectual Property Management Board (IPB).
Egerton University		Research Policy[9] that provides for dissemination of research findings and an Intellectual Property Rights Policy[10] approved in 2010.
Jomo Kenyatta University of Agriculture and Technology (JKUAT)	Digital Repository Policy[11]	Intellectual Property Policy (IPP);[12] created an Intellectual Property Office (IPO) under the Office of the Vice Chancellor for effective management of intellectual property and technology transfer.

Table 8.1 History of Open Access and Intellectual Property Related Policies at Kenyan Universities *(continued)*

University	Open Access Policy/ Repositories	IP Related Policy
Moi University		Research Policy published in 2012; also an Intellectual Property Policy and has created the Technology Transfer Office (TTO) under the Office of the DVC (Research and Extension) for the sensitization of the staff regarding intellectual property management, among other functions.

Institutional Performance in Intellectual Property Protection and Open Science

Analyses of the institutional-level policies on OA and IP rights coupled with data on the performance of universities in publishing in OA journals, as well as applications for patents, demonstrate a co-evolving trend in Open Science approaches and commercialization. While there have been an increasing number of organizations embracing OA, Kenya has also witnessed an upward trend in patenting at public universities from 2003 onward. From 2004 when the first IP policy was established at Moi University, nearly all universities and research institutes today have some form of IP framework. The establishment and adoption of these policies have led to a discernible upward trend in IP protection at all the universities from 2003 onward.

There has been a positive trend inclined toward OA over the last ten years, with Kenya being ranked second after South Africa in terms of the number of organizations with online repositories in Africa, accounting for fifteen percent of such organizations. Of six universities investigated in Kenya, three have had more than seventy-five percent of journal articles in their repositories openly accessible. As shown in Figure 8.3, public universities, represented by the University of Nairobi and Technical University of Kenya, have averaged below fifty percent on their OA publications (even though the University of

Nairobi reversed this trend in 2014). On the contrary, private universities such as Strathmore University and United States International University – Africa (USIU) have on average ninety percent of their journal publications available as OA (with USIU having all its journal publications available as OA). It is to be noted that OA policies and open institutional repositories are a recent phenomenon for public universities in Kenya. As Figure 8.1 (see page 179) has shown, these OA policies have only been adopted since 2012. On the contrary, the private Kenyan universities examined in this study are affiliated with American universities where openness has deepened over time.

Figure 8.2. Trends in Patent Filings at Kenyan Universities (2005–2013)

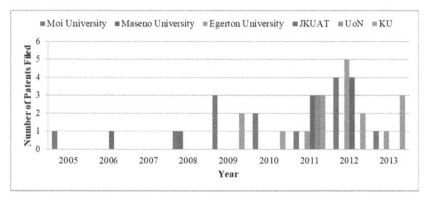

Figure 8.3. Open Access Articles from Selected Public and Private Universities (2005–2013)

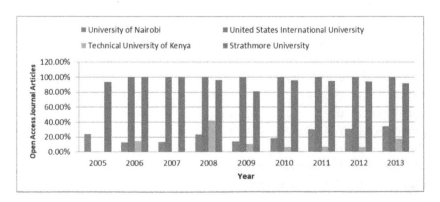

Discussion

Reflecting on the three case studies, the national and institutional policy context, as well as the empirical evidence on the co-evolution of patent applications and OA publishing, we frame our analysis around the following issues:

- Whether the existing "institutional regime" provides the necessary incentives for efficient creation and use of knowledge. In other words, do the existing policies, rules, and guidelines enable or hinder open, collaborative, and multi-disciplinary research partnerships?
- Whether the "governance structures" and patterns for both Open Science (as represented by OA) and commercialization (as represented by patent applications) support the researchers' quest to publish, innovate, and engage in further collaboration.
- The implications of both the institutional regime and governance patterns on the choices, behaviour, and practices of researchers involved in R&D collaborations.

The Institutional Regime, Open Science, and Collaborations

Our general observation is that (1) the national policy and legal environment is supportive of Open Science approaches, and government is encouraging increased openness in availability and access to information. Similarly, (2) openness is being embraced at the institutional level with universities adopting OA policies and establishing infrastructure to support wider dissemination of their research outputs.

Following North (1990), we define "institutions" to include both the rules (both formal and informal) and practices and their influence (as incentives or deterrents) in defining acceptable norms and behaviour of actors. More importantly is how this institutional regime affects the choices and practices of the actors. Kenya has put in place policies that favour openness in general and Open Science approaches in particular. Beginning with the country's supreme law, the Constitution, to its science, technology, and innovation policies and relevant sectoral laws and statutes, there exists a policy and legal framework to support Open Science. As already highlighted, Kenya's Constitution in articles 11, 33, and 35 not only recognizes the key role of science and technology in its development endeavours, but specifically pays tribute to the role of IP rights, as well as freedom of expression and access to information, including "academic freedom and freedom of scientific research." This overarching recognition in the supreme law is reflected in the country's development policies and plans. Dating back to the early 1980s when "Sessional Paper No. 5 of 1982 on Science and Technology for Development" was established, the role of scientific research and its relevance to development has been emphasized. This recognition has been carried on in subsequent policies culminating more recently with the STI policy in 2008 and the *STI Act*, 2013.

Similarly, the need for multi-disciplinary research partnerships has been recognized, and efforts have been instituted at the policy level to promote partnerships, especially with the private sector. "Sessional Paper No. 1 of 1994" and "Sessional Paper No. 2 of 1996 on Industrial Transformation" explicitly implore academic R&D institutions to forge links with the private sector. Following these policy provisions, a number of activities are being implemented to realize these objectives. For example, at the national level, this call to "openness" is punctuated with the government's Open Data project that seeks to make available all government information through a government open portal. Government agencies such as NACOSTI (and subsequently, the National Research Fund) are promoting collaborative, multi-disciplinary research partnerships through STI grant schemes. In order to qualify for these funds, research consortia must not only demonstrate a multi-disciplinary approach and team composition, they must also be multi-institutional and have private-sector partners in addition to respecting gender and other considerations.

This embrace at the national level is being replicated at the institutional level with universities and public research institutes establishing and adopting OA policies, open repositories for their research outputs, and recognizing publications in OA journals. Beginning in 2009, a number of universities established institutional OA policies and institutional repositories to share their research outputs widely and engage their local constituencies through activities such as Open Science week celebrations. Trends in OA publishing show that public universities have been embracing OA journals as a preferred channel of publication. This is supported by changing institutional policies so that they favour OA and universities putting in place the requisite infrastructure, including OA repositories, and sensitizing their staff on the need to embrace open publishing.

At the same time, there has been increased demand on universities and public research institutes to become more entrepreneurial and build linkages with the private sector, based on their research outputs. The rallying call has been for universities and public research institutes to produce research with commercial potential and interact more closely with the intended beneficiaries of their research. This too has pushed universities toward more IP protection for their research outputs. As in the case of institutional OA policies in publishing, there is an equal push for more IP protection in the institutional IP policies. The trends in the establishment of OA policies and institutional repositories, as well as the establishment of IP policies and infrastructures (depicted in Figure 8.1), paint a picture of co-evolution of IP and OA regimes. This co-evolution in policies is also reflected in their performance; there's a concomitant growth in both patent applications from the universities (Figure 8.2) as well as publications in OA journals (Figure 8.3).

Governance: At the Intersection of Policy and Practice

It is our consideration that the lack of guidance on IP ownership in research partnerships is a bombshell waiting to explode. This observation is borne out of a number of issues. First, while on the one hand there has been increased emphasis on the need to embrace collaborative interdisciplinary research, on the other hand there seems to be very little consideration to addressing issues of IP rights before, during, and after the research phase. For example, at the national level, policies and legal frameworks are supportive of Open Science, and nearly all public universities and research institutes in Kenya have developed institutional IP policies that define ownership and benefit sharing for inventions made by their staff and students. Similarly, there are publication guidelines (including OA policies) and copyright policies that define authorship. However, the main government funding and regulatory agency, NACOSTI,[13] lacks guidelines for its grantees on how to handle IP rights, publications rights, and future collaborations. Since NACOSTI provides the regulatory link between the national and the institutional-level actors, this policy vacuum is a key impediment and undermines effective governance of R&D collaborations in the country.

Secondly, most of the universities and public research institutes also have some form of governance infrastructure consisting of IPMOs or TTOs backed by IP management boards or committees. They have also established/adopted OA policies and created institutional repositories managed in most cases by librarians. The functions of these offices span from research coordination, to industry liaison, community outreach, knowledge dissemination, and general management of partnerships and collaborations. However, the major challenge is that most of these policies, offices, and infrastructure are institution-specific, and rarely is any consideration given to cases of intra-institutional collaboration. In most cases, these policies are silent on how to handle IP rights and publication procedures when research involves external collaborators, thus creating room for uncertainty on how to proceed when faced with the need to commercialize research products, disseminate outputs, or engage in other collaborations. Most IP policies simply state that publications or benefit sharing, in such cases, will be guided by research/funding contracts. When such contracts from funders are equally silent, confusion and uncertainty sets in.

Implications for Innovation, Publications, and Collaborations

From the case studies highlighted above, the common issue identified as an impediment to research collaborations is the lack of a binding framework on how to address IP issues. This has direct implications for innovation, publications, and future collaborations. In Case Study 1, the lack of policies on how partners would handle IP issues created a vacuum whereby the partners' failures to sign a consortium agreement and define IP ownership led to problems at the tail end of the research. Eventually, the private-sector partner took control of all the data and went ahead to generate new products from the research, locking out the public-sector partners.

In Case Study 2, the MSc and PhD students in the consortium are developing products and testing them for commercial viability. Management has adopted a "wait and see" approach to the IP issues. Should any of the products prove commercially viable, the stakes would increase and real conflicts would arise.

In Case Study 3, disagreements emerged over sharing resources (particularly budgetary allocations and duties), and left some partners feeling shortchanged. When the issues could not be resolved through internal mechanisms, one of the partners bolted but used the information and data in the original proposal to negotiate and enter into other collaborations. Even though the partners in Case Study 3 felt this was an infringement, they lacked the formal or legal avenue for redress since there was no binding document or guideline for addressing such issues.

Conclusion and Recommendations

In concluding, we revisit our central thesis: the lack of a guiding framework for negotiating and managing potential conflicts arising out of research partnerships presents a vacuum that may undermine open and collaborative research in Kenya. Such potential conflicts could arise over ownership of IP rights, publication and authorship rights, as well as data ownership and follow-on innovation. We sum up our observations by looking at the implications of this vacuum on researchers' choices, practices, and behaviour.

As can be gleaned from case studies, confusion, uncertainty, and unpredictability reign when there is no formal guidance to shape behaviour. This situation only helps to further erode trust among partners and undermine the goals of Open Science. From some partners taking control and locking out others as in Case Study 1, to some taking undue advantage and using collective resources/knowledge for personal gain through forging other partnerships as in Case Study 3, others have adopted a "take-no-action-till-it-happens" approach and are "waiting to see" what happens, preferring to deal with issues as they arise. Besides the lack of a binding institutional framework (in the form of rules and guidelines) to resolve conflicts, equally common in all these cases is a lack of a governance framework that spells out key roles and management responsibilities of the partners. While the proposal documents often spell out the "technical responsibilities" of each partner, only minimal considerations are given to "management and administrative responsibilities" as well as "conflict resolution mechanisms."

In our view, this is the gap that separates policy from practice. While there are policies at the national and institutional level to support open and collaborative science and commercialization, the behaviour at the actual project level is different; researchers are making choices and engaging in practices that serve to undermine the goals of open and collaborative science. This deviation of practice from policy arises partly from the fact that IP policies of the consortium members are "institution-specific" and do not have provisions for benefit sharing in case of "inter-institutional" collaborations. Each entity (whether a university, a research institute, or even an organization in the private sector) has their own internal policies that define how their staff should handle IP issues, how benefits are shared, and what kind of support they can access from their IP management offices, among other provisions. Similarly, the grant management policies mostly define how researchers can access funding from their grant offices. The challenge is that these policies are mostly "inward-looking"; that is, they only consider internal research processes and staff of the particular entities. When research involves other partners (who also have their own policies) and where partner policies conflict, there is no framework on how to resolve turf and supremacy contests that are likely to arise.

This policy vacuum calls for an overarching institutional and governance framework to guide choices, practices, and behavioural norms that promote trust and goodwill among partners. Such a framework will:

1. give researchers the much needed confidence to make their research findings openly accessible and the freedom to collaborate with other parties in pushing the research findings beyond the research shelves and into the market;
2. allow researchers to leverage other parties' strengths: for example, engaging the private sector for different kinds of support including financial, infrastructural, and experiential technical expertise; and
3. promote cross-institutional partnerships by defining key principles on "hierarchies, roles and responsibilities" that would help in negotiating and resolving conflicts.

Notes

1. See www.opendata.go.ke and www.icta.go.ke/kenya-open-data-initiative-kodi for more information.
2. See https://uonlibrary.uonbi.ac.ke/index.php?q=node/1482.
3. See http://erepository.uonbi.ac.ke/.
4. See https://goo.gl/dPDeFi.
5. Available at https://goo.gl/pnNwpk.
6. See https://goo.gl/uKN9mo.
7. Available at http://ir-library.ku.ac.ke/ir.
8. See https://goo.gl/kqnLhd.
9. See https://goo.gl/e7K4Ge.
10. Available at https://goo.gl/p8U3kQ.
11. Available https://goo.gl/8ikHdy.
12. See https://goo.gl/aiyPNf.

13. It is important to note that NACOSTI had been responsible for research funding at the time of this study, and the case studies were drawn from its database. However, following the enactment and operationalization of the *STI Act*, 2013, this function has now been taken over by the National Research Fund and NACOSTI retains the regulatory functions, quality assurance, and licensing.

References

Egerton University. 2014. *University Intellectual Property Rights Policy*. Egerton, Kenya: Division of Research and Extension, Egerton University.

Egerton University. 2014. *University Research Policy*. Egerton, Kenya: Divison of Research and Extension, Egerton University.

JKUAT (Jomo Kenyatta University of Agriculture and Technology). 2008a. *Intellectual Property Policy (IPP)*. Nairobi: JKUAT.

——. 2008b. *Digital Repository Policy (IPP)*. Nairobi: JKUAT.

Kenyatta University. n.d.-a *Intellectual Property Policy*. Nairobi: Kenyatta University.

——. n.d.-b *Kenyatta University Open Access Institutional Repository Policy*. Nairobi: Kenyatta University.

Moi University Research Committee of Senate. 2011. *Moi University Research Policy*. Eldoret, Kenya: Moi University. https://research.mu.ac.ke/wp -content/uploads/2018/06/Research-Policy-May-2010.pdf.

North, Douglass C. 1990. *Institutions, Institutional Change, and Economic Performance: The Political Economy of Institutions and Decisions*. New York: Cambridge University Press.

Republic of Kenya. 1982. "Sessional Paper No. 5 of 1982 on Science and Technology for Development." Nairobi: Government Press.

——. 1994. "Sessional Paper No. 1 of 1994 on Recovery and Sustainable Development to the Year 2010." Nairobi: Government Press.

——. 1996. "Sessional Paper No. 2 of 1996 on Industrial Transformation to theYear 2020." Nairobi: Government Press.

——. 2007. "Kenya Vision 2030." Nairobi: Government Printers.

——. 2008. "Science, Technology and Innovation Policy and Strategy." Nairobi: Government Printers.

Republic of Kenya. 2010. "The Constitution of Kenya." Nairobi: National Council for Law Reporting. http://kenyalaw.org/kl/index.php?id=398.

——. 2012. "Sessional Paper No. 10 of 2012 on Kenya Vision 2030." Nairobi: Government Printers.

UoN (University of Nairobi). 2013a. *Intellectual Property Policy*. Nairobi: University of Nairobi.

——. 2013b. *Research Policy, Abridged Edition*. Nairobi: University of Nairobi.

UoN. 2014. *Open Access Policy*. Nairobi: University of Nairobi. https://uonlibrary .uonbi.ac.ke/index.php?q=node/1482.

UoN. 2016. University of Nairobi Digital Repository. Nairobi: University of Nairobi. http://erepository.unobi.ac.ke/.

SECTION 3

NEGOTIATING OPEN SCIENCE

INTRODUCTION

Hebe Vessuri

The three chapters in this section, which I will distinguish by their author organizations—Natural Justice (Chapter 9), STEPS (Chapter 10), and CONICET (Chapter 11)—are concerned with the intricacies of negotiating openness, knowledge, and research procedures, including the definition of the very problems to be investigated in open collaborative research aimed at producing useful knowledge. I would like to highlight some of the areas of discussion that have usually been neglected or underemphasized in discussions of Science Openness and which are taken up in the three papers. Each deals with specific aspects of the problem.

The particular contributions made by each add to the others and could eventually become the building blocks of a single, combined approach to scientific research in a new key. Thus, Natural Justice describes the challenges of negotiating research contracts between researchers and Indigenous communities in truly collaborative projects, where research questions would flow from the needs and interests of Indigenous peoples and where academics, non-profit researchers, and Indigenous peoples would be equal partners in the production of knowledge. STEPS explores what the best spaces and strategies are to start the process of Open Science: the tools and capacities that need to be developed and the challenges faced by practising Open Science in different contexts. CONICET, in turn, more generally aims to investigate the conditions under which scientific knowledge produced in varied regimes of openness, in different contexts, and with diverse actors has the capacity to be used in order to deal with—or even resolve—social problems.

Criteria

CONICET's criteria for selecting the case studies were: the kinds of knowledge involved; how heterogeneous stakeholders intervene in the processes of co-producing knowledge and public problems; and feasibility and access to data sources. Since STEPS placed emphasis on policies for implementing Open Science practices, their criteria were how scientists are building their networks, what resources are available, who participates in Open Science practices, and what types of data are available or should become available in a country such as Argentina. Natural Justice's criteria were related to the multiple scales of histories, geographies, institutions, and ways of knowing involved in engaging in an open and collaborative research project.

Negotiating Openness

Open Science appears not as a simple, neutral notion, but as a complex array of decisions and distance taking by scientists and non-scientists alike, with moving boundaries pragmatically kept. For Natural Justice, openness is not an end in itself, but involves recognizing potential downsides, especially if only some elements of openness are asserted in a unilateral, exploitative, and partial fashion. The study argues that a collaborative project like this one requires a more "situated" approach to openness, and it flatly rejects the notions of science as "open" and nature as "freely accessible" for having been historically invoked to exploit countries such as South Africa: "The notion that knowledge and resources should be open and accessible has been historically misused to cast countries in the Global South, including South Africa, as suppliers rather than producers of knowledge, and in particular Indigenous peoples' knowledge, resources, and heritage as free for the taking."

STEPS tries to understand how openness is realized in the situated context of Argentina. It analyzes the characteristics and scope of openness—how it has been opened (participation and barriers on access) and who is involved in the processes of openness (for whom it is opened and for which uses and benefits). The four cases explored have the common goal of opening data for re-use in scientific networks and by citizens, although they have different results. The chapter suggests that the negotiation of the opening process is similar to the construction of boundary objects. There is a brief exploration of how scientists build boundary objects to negotiate the opening at three levels: tools and infrastructure, the opening of data to other experts (negotiating different meanings and uses of the data with new potential users), and in the communication and dissemination of results.

CONICET investigates the conditions under which scientific knowledge (produced in a more or less open way according to each particular case) is capable of being utilized to satisfy social needs. Their approach takes the social use of knowledge as its focus, not "subsequent" to its production but co-produced with it. With a sociological approach, focus is on the relationship between use of knowledge and public issues in non-hegemonic contexts, considering the configuration of public issues as both a social and cognitive problem. As argued, a given "scientific" definition of a problem puts forward certain specific views and solutions as "possible" and excludes others; the frameworks set by scientific knowledge, far from being universal, establish specific links between the problem in question and the different actors that mobilize it or are excluded from it. It is proposed that there are other requirements related to tacit knowledge and to social and political skills that stand in the way of effectively using openly accessible knowledge. Processes with greater collaboration in the production of knowledge do not imply an a priori determination of its effective use oriented toward satisfying social needs.

The Role of Drivers

The role of a certain type of actor—the "driver"—is significant. For CONICET, it is an actor who in some way marks an initiation or rupture, mobilizing scientific knowledge in a particular way in pursuit of a particular social use, and is also a highly active and influential element in shaping the public problem. According to STEPS, in the case of Open Science it is still not clear who will push for this idea and how scientists are going to engage in the process. It considers scientists who learn to negotiate their interests and practices during the opening process. In particular, it argues that the further scientists engage in the opening process, the more capabilities and tools they will need, though none of which is currently being provided by scientific institutions or policy schemes. Policy makers might need to consider better policy design to promote Open Science. Through the construction of boundary objects, scientists are introduced to new fields: (1) a relational field that allows interaction with scientists in other disciplines and with the general public; (2) a technological field that facilitates the development and use of new open technologies; and (3) a management field, which shows that when engaging in Open Science practices, there some difficulties in collaboration remain across different disciplines in the project.

Material Dimension of Knowledge

The material dimension of knowledge is significant in relation to the possible forms of "use" and "openness." CONICET explores ways in which knowledge is made utilizable. In Chagas disease research, it is scientific publications, or rather information outputs codified and organized into databases (DNA sequences); in the Jáchal-Veladero case, technical reports do not operate on the material form of knowledge, but on its socio-cognitive content and on its problematic criteria of elaboration; the strategies for conserving threatened species, and the cognitive problems of a discipline whose empirical objects are distributed on a wide-ranging regional scale mean a greater possibility of openness, both in terms of the use of technological infrastructures and of human collaborators. The social sciences make the boundaries between knowledge producers and the data-providing subjects more nebulous; the frontiers between the different disciplines (anthropology, history, and sociology) are less clearly demarcated than in the "hard" sciences, allowing for varying degrees of integration.

Natural Justice describes the challenges of empowering indigenous peoples and knowledge systems in connection with climate change and intellectual property rights. The chapter describes the process of negotiating research contracts with Indigenous communities and how they conceptualized the concept of a "situated openness" as they became more familiar with both the different and similar traditions of producing and disseminating their knowledge. This helped them understand the relations of power that enable or hinder open and collaborative research.

The cases chosen by STEPS belong to different networks of knowledge production: astronomy; biology, limnology, and climate change; and ornithology and chemistry, geography, and history, which have been relatively successful in opening at least one of the research phases. STEPS makes the useful point that their researchers do not normally commit to total openness, but rather attempt to open up pragmatically. However, it is still not clear what aspects of the research cycle scientists and institutions choose to open and how — what negotiations take place? Natural Justice, on the other hand, is concerned with knowledge that is the intellectual property of Indigenous communities. It is interesting to observe that in both cases opening up is seen pragmatically.

Public Knowledge

Natural Justice analyses the particular challenges in the notion of public knowledge. It deals with community knowledge, which leads the authors to consider the differences between these two notions. They observe limitations in community research contracts. Although the contractual provisions are meant to disrupt hierarchies between researchers and the researched, it is unclear if contracts are the appropriate vehicle for reducing hierarchies of knowledge production. Only those who sign the contracts are bound by them for the specified duration, which means that third parties having access to the indigenous knowledge (IK) are not bound by the responsibilities set out in the contracts. This is a convoluted way of showing that public knowledge does have its positive and not so positive sides.

For CONICET, public problems are processes in which unequally distributed resources become mobilized. Strengthening and institutionalizing public forums could be a way to foster the mobilization and production of knowledge aimed at addressing social needs and demands. The challenge here lies in ensuring legal state support while, at the same time, enabling local stakeholders to retain their autonomy against potential mechanisms of co-optation induced by political, scientific, or economic corporatism.

Although the four cases in the STEPS chapter have implemented some form of Open Access, thus eventually allowing data to be reused by other scientists, there is little evidence that this is happening at the local level, in contrast with international cases. There are still some difficulties with collaborating across different disciplines involved in the project.

Citizen Science

CONICET's proposal is partially amenable to the ideas of citizen science, in the sense that it calls for a systematic fostering of institutional spaces for both scientific openness and political participation. However, it also goes beyond citizen science as it understands that scientific knowledge and public processes co-produce each other, rather than being just knowledge outputs that "inform" political decision-making.

Natural Justice reflects on tensions related to openness in research in collaboration with Indigenous peoples regarding their knowledge systems and intellectual property rights, the importance of considering contexts in which the current research is located, and that Open Science practitioners need to acknowledge injustices faced by Indigenous communities both historically and in the present day. Problems are compounded; for instance, ethics approval processes that are based upon the notion that knowledge is individually held will not meet the needs of many Indigenous communities who view their knowledge as being collectively held.

One of the case studies in the STEPS chapter considers the experience of a group of scientists and students from the Laboratory of Research and Formation of Advanced Informatics from the National University of La Plata who have started a citizen science initiative using NOVA Open Data. Specifically, they have begun to develop electronic games that allow the general public to collaborate in the classification of data, such as of galaxies. In another case study, the Integrated Land Management Project researchers are shown to be cautious regarding the management of neighbours' expectations since they cannot guarantee that solutions will actually take place. On their side, the neighbours are also cautious about their degree of commitment to the project; this was not the first project that had required their collaboration but did not always deliver the expected solutions. On the other hand, e-Bird is a citizen science project that receives bird sightings from anybody in any part of the world through a website and mobile phone applications launched in Argentina by an NGO with the support of a network of eighty bird watching clubs (*Clubes de Observadores de Aves — COAs*).

Concluding Remarks

The enormous variation and diversity of situations made visible by the individual studies considered can be gauged by the complexities and richness of the processes of negotiation that take place when different social actors, having different views, interests, and power, engage in joint efforts to define and solve a social problem. By the same token, they reveal that other combinations, emphasizing other similarities and differences, would have been possible. The factors involved are seen differently by the different partners, although they would eventually agree on their priority. Clarification is needed at different stages in the negotiation process to avoid misunderstandings which may cause problems and create barriers to reaching beneficial outcomes. Negotiation skills are required for a wide range of activities. Negotiation implies the reconciliation of multiple views and opinions; it takes time to arrive at a group decision and the co-construction of knowledge. Negotiation can also be seen as the intertwining of perspectives, contributed by the different social actors, and the merging of these into a common shared perspective.

We appreciate the centrality of negotiation within each of the different frameworks developed by the papers in this section. The following are shared features: small group processes; social constructivism; a search for shared understanding of the knowledge object; and the distributed, problem-based learning whereby the group negotiates lists of problem statements, key evidence, and working issues. There is negotiation and re-negotiation of the group's understanding throughout the learning process, leading eventually to distributed cognition. Knowledge is frequently distributed among the abilities of group members and the artifacts that they use. Accordingly, it is co-constructed by interactions among people and their shared artifacts, including prominently by means of negotiation practices that result in establishing a common ground for understanding. The three studies emphasize the exploration of bottom-up processes that often go invisible or get lost when they are absorbed in larger structures.

CHAPTER 9

Co-production of Knowledge, Degrees of Openness, and Utility of Science in Non-hegemonic Countries

Hugo Ferpozzi, Juan Layna, Emiliano Martín Valdez, Leandro Rodríguez Medina, and Pablo Kreimer

Abstract

Collaboration in scientific knowledge production has been histori-cally dominated and driven by hegemonic (Northern) countries, while non-hegemonic countries tend to take on secondary roles. The growing discourse on Open Science provides the opportunity to look critically at the roles and outcomes of collaborative knowledge creation. Drawing on four diverse case studies throughout Latin America, this project has sought to assess the ways that diverse actors, processes, and sectors converge to collaborate (willingly or not) on resolving social issues. Using Open Science as a theoretical framework, the chapter concludes with a summary of how different "types" of challenges may be more or less amenable to the collaborative practices of Open Science.

Introduction

The general orientation of this chapter is to investigate under what con-ditions scientific knowledge, produced in varied regimes of openness in different contexts and with the participation of diverse actors, can be utilized to address, and perhaps even resolve, social problems. With that aim, we use Open Science as a theoretical framework that, within the social studies of science, mobilizes different concepts which are normally considered separately, and that enable us to take some steps toward constructing a more comprehensive and integral approach to openness. These concepts have been operationalized in the study of four empirical cases that relate to distinct configurations of knowledge, actors, contexts, institutions, and regimes of openness. The case studies are:

1. national and international networks dedicated to Chagas disease research;
2. disputes about the environmental contamination of a mine in the Andes mountain range;
3. strategies for the detection and conservation of jaguars in the tropical forests of northeastern Argentina; and
4. the production of social science knowledge on North-South migrations in Mexico.

The selection of these cases was made on the basis of three criteria. The first criterion focused on the types of knowledge and disciplines involved, which are very different in each of the four cases: basic knowledge in Chagas disease research and applied knowledge in the cases of wildlife preservation, both within the life sciences. The case of migration studies, on the other hand, belongs to social sciences; and last, the case of mining disputes integrates all of the former within a space of political controversy. Second, these cases explore how heterogeneous stakeholders intervene in the application of knowledge by examining different processes of knowledge co-production geared toward addressing public problems. Third, the cases were also selected on the basis of feasibility and access to data sources. The diversity of knowledge and types of stakeholders discussed might help in clarifying the conceptual tools proposed to understand openness and uses of knowledge.

Taking into account the emergent elements of these four case studies, toward the end of this chapter we suggest a preliminary typology with which to systematize the most significant dimensions in the regimes of knowledge openness and the possibility of using knowledge to address social needs in non-hegemonic contexts.

We focus on three central problems crossing the processes of production and use of scientific knowledge in non-hegemonic contexts (Losego and Arvanitis 2008).

Firstly, we consider the historical problems facing Latin American societies in relation to putting locally produced scientific knowledge to effective use. Indeed, these difficulties were identified in the 1960s, and various analyses and policy alternatives have been put forward. Thus, Sábato and Botana (1969) proposed to analyze the relationships between the production and use of knowledge with the well-known formulation of a triangle of relations with scientific-technological infrastructure, government, and the productive sector at each vertex. It was noted that, while the links between academia, business, and the government were fluid, their reciprocal relationships were very weak or non-existent, so that policy efforts should be oriented toward designing instruments to promote stronger links.

From the 1980s onward, several mechanisms were implemented to stimulate "university-productive sector" relations (Sutz 1994; Arocena and Sutz 2001), what have been described as "linking" policies. Over the following decades, while these relationships were formulated in similar terms, they were also connected to the idea of a "triple helix," in which the axes are the same as the triangle presented by Etzkowitz and Leydesdorff (2000). Similar considerations were posed in terms of a "national system of innovation," whose formulation is due to the well-known book edited by Lundvall in 1992.

These ideas were adopted rather uncritically by several studies in Latin America and other developing regions, and, above all, by policy makers (IADB 2001; STI Law 25.467 in Argentina, etc.). The common problem in all these approaches was their uncritical approach toward the modes of knowledge production: science is taken as a "commodity" or object to be "transferred" (generally neutral in content), and the goal was to locate the main problem in finding better mechanisms for its transfer from one context to another.

We have contested these perspectives for several years (Kreimer 2003; Kreimer 2014), arguing that the social use of knowledge is not something that is found "at the end of an assembly line," as a recreation of a linear model, but that it should rather be understood as a more complex process in which the utility of knowledge informs the very processes of scientific research. We have drawn the conclusion that a hallmark of developing countries is precisely the difficulty of being able to effectively use locally produced knowledge, whether to address social-environmental problems or to contribute to industrial and social development. We identified this process as AKNA: Applicable Knowledge Not Applied (Kreimer and Thomas 2005).

Our research focuses on a second problem: the relatively peripheral position of Latin American countries. It has been evident that peripheral regions faced serious obstacles to their scientific development in relation to the universalization of science. As Losego and Arvanitis (2008) point out, "Non-hegemonic countries are dominated in the international division of scientific work" (343). This idea is already present in the concept of peripheral science: scientists do participate in international collaborations, but are frequently undertaking "secondary" functions or subordinate work in programs elaborated in hegemonic countries (Díaz, Texera, and Vessuri 1983; Kreimer 2014).

In this sense, two aspects converge to hinder the use of knowledge in developing countries: the role of the scientific elites and the prevailing evaluation systems that, with a few exceptions, tend to prioritize the publication of articles in high-impact international journals, whose agendas are markedly dominated by the issues and methods which interest the great powers. In turn, these elites are increasingly co-opted to work on projects with international cooperation in which they undertake relevant activities that nonetheless have a high technical content and little leeway to develop theoretical concepts. In this way, cognitive control is exercised by hegemonic groups and research centres on a process dominated by a sharp division of labour and a logic of subordinated integration (Kreimer and Levin 2013). In addition, although in "North-South" international collaborations it is possible to industrialize the knowledge generated collectively, the companies located in the hegemonic countries are usually responsible for doing it.

A third issue relates to the modes of "openness" or "closure" of the processes of scientific research. The polysemic concept of Open Science functions as a wide umbrella. In this sense, it is worth revisiting the classification, including the five schools of Open Science advanced by Fecher and Friesike (2014), who consider:

(1) the infrastructure school, concerned with the technological architecture; (2) the public school, concerned with the accessibility of knowledge creation; (3) the measurement school, concerned with alternative impact measurement; (4) the democratic school, concerned with access to knowledge; and (5) the pragmatic school, concerned with collaborative research.

Each of these approaches places emphasis on different relational aspects, but we wish to concentrate particularly on the "pragmatic" school (although the label is not entirely convincing), and also on the "public" school of Open Science, even though we have to refer to the infrastructure school as well (concerned with the material platforms that support knowledge).

Our sociological approach is concerned with the question about the actors who participate in the processes of production and use of scientific knowledge, linked to the idea of co-production proposed several years ago by Jasanoff (2004), who suggests that:

> Knowledge and its material embodiments are at once products of social work and constitutive of forms of social life; society cannot function without knowledge any more than knowledge can exist without appropriate social supports. Scientific knowledge, in particular, is not a transcendent mirror of reality. It both embeds and is embedded in social practices, identities, norms, conventions, discourses, instruments and institutions… .(3–4).

We add to this idea of co-production a central concern that not only is knowledge "co-produced" but its social uses are actually inscribed in the processes of production themselves, in which the role of different actors is crucial. This approach enables us to go beyond formal openness and to focus on the relationship between use of knowledge and public issues in non-hegemonic contexts. We consider the configuration of public issues as both social and cognitive problems. A given "scientific" definition of a problem puts forward certain specific views and solutions as "possible" and excludes others. The frameworks set by scientific knowledge, far from being universal, establish specific links between a given problem and the different actors who mobilize it or are excluded from it.

From this perspective, even open processes of knowledge production cannot ensure that knowledge will be a priori oriented toward satisfying social needs. Indeed, the very definition of the "scientific problem" plays a crucial role in the public arena, as it sets the different instances through which knowledge is transformed, used, and implemented. In turn, this perspective allows a deeper understanding of the social and cognitive barriers frequently dismissed by other approaches to Open Science. Apart from the material and formal requirements, we propose that there are other requirements related to tacit knowledge and to social and political skills that stand in the way of effectively using openly accessible knowledge. Cognitive barriers, then, entail sophisticated knowledge or technical requirements that cannot be fulfilled by all the concerned stakeholders. However, the boundary between strictly cognitive and other kinds of barriers is rarely clear-cut, as the production and use of scientific knowledge must often be accompanied by legal, political, or interactional knowledge about how its potential users or audiences can be addressed or enrolled.

In this way, we explore to what extent the fact that diverse actors participate, even in a controversial way, in the material and symbolic processes of production of cognitive objects has an influence on the uses of knowledge. From this perspective, processes with greater collaboration in the production of knowledge do not imply an a priori determination of its effective use oriented toward satisfying social needs. The three perspectives presented here should be considered together in order to furnish us with an integral image of the different dimensions related to the degrees of openness of knowledge, their actual or potential uses, and the broadest contexts in which these processes take place in a globalized world.

Empirical Case 1: Chagas Disease Research and its Networks of Knowledge Production

Chagas disease is endemic in Latin America, affecting around ten million individuals. As a consequence of recent migratory processes, the disease has also spread to non-endemic regions, although it has only recently become an actual public health issue (i.e., Hotez et al. 2013). Known as *American Trypanosomiasis*, it is mainly transmitted through the bite of insect vectors called "kissing bugs" or "*vinchucas.*" These bugs inhabit rural households across the Americas and introduce the *Trypanosomacruzi* (the parasite that causes the disease) into the host organism after feeding on their blood. During the chronic phase ensuing infection, the disease causes cardiac and gastroenterological disorders. In view of its epidemiological patterns and the lack of an effective treatment for it, the World Health Organization (WHO) has classified Chagas in the group of seventeen neglected tropical diseases (WHO 2012).

The advances in biological research into *T. cruzi* in the 1970s reinforced local and international scientific interest in the disease, drawing the attention of global health organizations and research centres such as the WHO's Special Programme for Research and Training in Tropical Diseases. In the 1990s, the causing organism was part of the *T. cruzi* Genome Project (TcGP), an internationally collaborative initiative aimed at sequencing its genome, which spanned more than a decade. Doctors Without Borders and the Bill & Melinda Gates

Foundation are other international institutions interested in research into the disease and its potential eradication.

Even with sustained support for biomedical research and its focus on the potential development of therapeutic applications, so far there is no effective treatment for Chagas disease: the only drug currently used, whose effectiveness is still limited, was developed fifty years ago by Roche. Hope for the development of suitable drugs was placed—similarly to the Human Genome Project—in the TcGP and the genome databases developed afterward. One of the most striking examples is TDR Targets, an open genomics resource oriented toward prioritizing possible targets to attack the parasite using chemical compounds (Agüero et al. 2008; Magarinos et al. 2012; WHO 2007). Due to its Open Access resources and its potential for medical applications, we conjectured that the findings of the research into Chagas disease could be subjected to processes of cognitive exploitation. These processes imply the appropriation of knowledge by private actors without objective compensation for the producers. In this way, pharmaceutical firms could potentially take advantage of the research, which is basically financed by public funds and NGOs, in order to industrialize knowledge in the form of medical treatments that would otherwise not be profitable.

On the contrary, the possibility of developing applicable knowledge, sensitive to local needs, does not only depend on the production of and access to Open Data, but on a group of contextual interactions between the political and scientific spheres, as well as on the connections between public health, the affected populations, and the private companies in charge of the development of treatments. In effect, the path to implementing the commercialization or distribution of a drug is slow and difficult; it normally requires dealing with government offices in different jurisdictions, negotiating the prevailing legislation, carrying out reliable clinical trials, and, last but not least, making its delivery viable in economic terms (Masum and Harris 2011; Porrás et al. 2015).

The inadequacy of the more restricted notions of access and openness also emerge upon examining the knowledge production about the disease in the fields of biomedicine and genomics. In recent decades, representatives from these fields became spokespeople for the issue, and biomedical research was conceived, in itself, as a "legitimate" strategy for intervening in the problem of Chagas disease. However, this highly internationalized production of scientific knowledge makes it difficult for those affected to participate in the formulation of the problem at stake and the decisions connected to research (Kreimer 2015; Kreimer and Zabala, 2007). The dynamics of knowledge production, furthermore, are highly dependent on the institutional, symbolic, and material support provided by global NGOs and research centres in developed countries. As the literature has shown, the interests of this group of global biomedical actors barely contemplate the particular needs of the local contexts where they act (Behague et al. 2009; Leys Stepan 2011).

Lastly, the limitations of the classic concepts of access and openness can also be observed among the researchers and health professionals themselves. The professionals engaged in patient care are, in general, detached from the production of knowledge and decision-making regarding research, and their capacity to access resources is significantly less than those in the biomedical field (e.g., Sosa-Estani 2011).

Empirical Case 2: Socio-technical Dispute Around the Cyanide Spill in Veladero, Jáchal, San Juan

The Veladero mine in San Juan province extracts and processes gold and silver by means of "cyanide leaching," also known as opencast mining. In September 2015, thanks to a Veladero employee, the news of a cyanide solution spill into the watercourse, which feeds the rivers vital for the mine's neighbouring communities, circulated unofficially in social networks. Rapidly, several officials from the Ministry of Environment described the event in the media as an "environmental incident," thus defining the public problem (Gusfield 1981) that is at the centre of the dispute analyzed here. In this dispute, the production and mobilization of knowledge played an important role in achieving more mediate objectives. Briefly, the sectors in conflict are, on the one hand, a block whose most prominent actors are the provincial executive power and the Barrick company, along with some media outlets, public and private universities, environmental management institutions, and business groups related to mining. On the other hand, there are the "Hands Off Jáchal" Assembly (*Asamblea de Jáchal No Se Toca"—AJNST*) from the homonymous city, along with various organizations engaged in environmental struggles. This last group demands the immediate closure of the mine.

A significant feature of this conflict is the production of technical reports done privately in payment for services. This is enabled by a modality of university-industry collaboration based on the notion of "transfer" (OECD 1996) as universally useful and, therefore, freely commercializable knowledge. These studies are characteristically secret, both in their elaboration process and in consumption, which comes under the absolute authority of the "purchaser." Another factor must be added: territorial control of the mine, and, therefore, of the object of study itself, is located within the exclusive control of the Barrick corporation.

Regarding how to approach and resolve the public problem, we find a very particular configuration: a judicial ruling is processed through private reports and elaborated in highly restricted conditions. This is of great importance, not only to understand the character of the dispute and the conditions under which it developed, but also to more concretely approach the aspect related to the social uses of the knowledge. Following the spill, several officials from the provincial executive power issued to the press the findings of various technical reports commissioned by different institutions (UNSJ, OSSE, and others), all in one way or another linked to the provincial government. All these reports indicated normal, or, even in some cases, nonexistent, levels of cyanide, with no reference to any other type of potentially toxic substance. The outcome was predictable: the continuation of Veladero's operations without major disruptions.

However, the AJNST successfully undertook various procedures through its political organization, reinforced by mass participation. Firstly, via a demand made to Jáchal's mayor, it was able to mobilize the laboratory at the National University of Cuyo in the province of Mendoza, especially selected due to its location beyond the sphere of influence of the San Juan executive power. The findings of the report made by this laboratory did reveal the presence of cyanide, but mainly found concentrations of heavy metals that made the water unsuitable for human consumption.

Then, opening up a new political and cognitive stage in the dispute, the AJNST drew the national judicial power into the dispute by means of a petition against Barrick and state officials for committing infractions affecting interprovincial or national watercourses. In February 2016, the federal court ordered new studies from other institutions that produced results agreeing with those the University of Cuyo published in September/October 2015. Afterward, the federal court facilitated the intervention of Robert Moran, an internationally renowned hydrogeological mining expert, chosen and enlisted by AJNST. Moran was the first representative of AJNST's interests to enter Veladero to determine the character of the events surrounding the spill. Through Moran's participation, the AJNST managed to gain access to the actual conditions of knowledge production. This meant significant progress by AJNST with regard to the capacity to produce new technical knowledge and also to challenge those elaborated by sectors connected to the company.

This leads our analysis to various observations. Firstly, the degrees of access to knowledge enjoyed by AJNST changed throughout the different stages of the dispute. Additionally, this was a process interdependent of the development of AJNST's socio-cognitive skills, which included reasoning about diverse technical problems, the ability to define cognitive criteria, mobilizing university laboratories, and choosing and enlisting national and international scientific actors. Lastly, the changes in the extent of access to knowledge and the recognition that these skills are co-produced (Jasanoff 2004) and, in turn, along with an equally dynamic and changing aspect, political-organizational skills become visible in the constitution of the Assembly itself, as well as in the political alliances forged with diverse groups.

Empirical Case 3: Collaborative Jaguar Monitoring Networks

The *yaguareté* (in *Guarani*), or jaguar, is the largest feline in the Americas and the third largest feline species in the world. Despite its conservation status being variable due to its wide-ranging distribution across the continent, it is considered in Argentina to be under threat of extinction (Ojeda, Chillo, and Diaz Isenrath 2012). Currently, the jaguars found in this country are distributed as three subpopulations in the Yungas (Jujuy), Chaco, and Misiones.

In Misiones, the subpopulation is isolated and has suffered a reduction in numbers over the last twenty-five years of between two and 7.5 times its population density (Paviolo et al. 2008). The first studies of the jaguar in Misiones date back to 1990 and 1995 and were carried out by Peter Crawshaw (Crawshaw 1995). Crawshaw's work is highly valuable, even though his estimates are not precise. His principal working method consisted of capturing specimens and fitting them with collars with a radio-signal transmitter, and triangulating to establish the source of the signal. This technique is known as radio telemetry (Di Bitteti 2015).

In 2002, the Argentine Wildlife Fund, a conservation NGO, kick-started the initiative to advance knowledge about jaguar populations in the Alto Paraná Atlantic Forest, aware of the need for information to validate conservation action plans for the jaguar in Misiones. In this context, the "Yaguareté Project" (2002–2016) is the result of a collection of scientific research produced by the IBS-Conicet Ecology and Mammal Conservation Group located in the city of Puerto Iguazú, Misiones (North East Argentina, close to the Brazilian border). The initial goal of the project was to assess the conservation status of the jaguar and puma populations in the Alto Paraná Atlantic Forest region and to identify their main threats (Di Bitteti 2015).[1]

The cognitive problem hides a series of practical problems that are very difficult for a "traditional" scientific organization to solve. On the one hand, there is a team of three researchers with limited funds, needing to collect data over an extended time span; on the other hand, there are two nocturnal animal species that live in low densities distributed over a hard-to-access geographical area without communication infrastructure—more than twenty-seven million hectares distributed in three countries (Argentina, Brazil, and Paraguay).

To determine the presence of these species, a participatory network of volunteers and collaborators was established with researchers from the three countries that share the Alto Paraná Atlantic Forest area (De Angelo et al. 2011). Participants were trained in simple methods of collecting big cat fecal samples and footprints (indirect methods of detection), and between 2002 and 2008, more than three hundred volunteers helped obtain 1,633 records of pumas and jaguars. Inscribed in the field of conservation biology as a discipline, the first thing that springs to our attention is that "biodiversity" (the main objective of this field) as a discrete reality composed of an infinite number of living beings (including plants, animals, microorganisms, humans, and their interactions) is unevenly distributed over geographical space.

Starting from the principle outlined by Whitley (2012) in relation to the structure of knowledge issues influencing the social organization of science, it is possible to consider that due to "biodiversity," as the main physical reference point of the research questions and the problems of conservation biology being "distributed" in the same way as the scientific collaborations, the putting-into-practice of citizen scientists' activities as a form of resolving problems of knowledge that are limited to a local or regional scale is more feasible in these types of disciplines.

In the last instance, disciplinary attachment is an important element to take into consideration since it allows us to see the disciplines as "loci" from which greater degrees of permeability are found (or not) with regard to the possibility of integrating citizen scientist practices into their core.

Empirical Case 4: Development of Social Science Knowledge In Relation to North-South Migrations in Mexico

As a public issue, migration in Mexico illustrates the three problems of knowledge production in non-hegemonic contexts. First, it shows that locally produced knowledge is not easily appropriated by locals. With few exceptions, results do not necessarily inform public policies (CIDH 2016; Calvillo 2015). Second, Mexican research on migration is permeated by the features of peripheral science and its tension between local relevance and international impact (Alatas 2003). Third, research on migration illustrates the co-production of knowledge by emphasizing how actors in different parts of the country problematize the phenomenon and, consequently, propose different actions to implement.

From an academic perspective, Colegio de la Frontera Norte has played a central role, thanks to its Survey of Migration at Mexico's northern and southern borders. Initiated in 1993, it attracted government offices such as Consejo Nacional de Población, Secretaría de Trabajo y Previsión Social, Instituto Nacional de Migración, Secretaría de Relaciones Exteriores, Secretaría de Salud Pública, Consejo Nacional para la Prevención de la Discriminación, and Secretaría de Desarrollo Social. It is practically impossible to find a project that is more articulated between academia and the governmental sphere or one that enjoys such support at the highest bureaucratic level. The survey is published annually, and its results are available to the public through its website and databases, in SPSS format, being opened up for direct consultation by interested parties. After twenty-four years, this continues to be a priority project, but it has also become an attraction for foreign graduate students who, as grant-holders, join this institution with the aim of taking advantage of the accumulated statistical data. Surprisingly, COLEF's survey is not formally associated with migrant non-governmental organizations. However, academics and students working on it have close contacts with these organizations since the identification of the "right" places to survey is knowledge accumulated by NGOs because of their presence in the field. In this context, it is evident that while the institution and its academics seem to share an interest in opening up the data and findings, the rules of the academic game that undermine a more integral form of participation by civil society actors, especially those directly involved, still prevail. In terms of co-construction of knowledge and public problems, migration in the north seems to be an issue that requires the involvement of the state at the highest level and of prestigious academic institutions since the results not only contain information for local actors but also data for political exchange and coordination with the United States. The country's asymmetry could be seen as a factor of pressure for COLEF and associates toward mainstream, "big" social science projects such as this annual survey.

On the southern border, the situation is also complex. Institutions such as El Colegio de la Frontera Sur, Centro de Estudios Superiores de México y Centroamérica (CESMECA), Centro de Investigación y Estudios Superiores en Antropología Social, and Región Sureste conduct research on the border area (whether directly or indirectly linked to migration). Unlike Tijuana, in the south the migratory phenomenon seems to be conceived as, among other things, impossible to approach other than through direct and permanent contact with civil society. As an interviewee put it, "civil organizations can give you the data quickly because they work directly with people, and there is a different way of data production, without intermediation" (interview 9, passage 1).

Given the conflict-ridden presence of the state and federal governments in the region, the Indigenous ethnic question, which complicates the panorama, and the lack of comparable resources in relation to institutions from other parts of the country, research into migration (and other areas) at the southern border seems more responsive to the specific needs, interests, and realities of local actors, particularly through NGOs and social movements. Similar to the north, the border here is not merely a research problem, but a situatedness that irredeemably puts scientists in contact with the subjects that experience and survive it. Unlike the north, the research is more intimately connected with the social subjects and only indirectly with the state and decision-makers. Migration in the south is co-constructed differently, including the asymmetry with Belize and Guatemala's academic communities. Perhaps as a consequence, an interest in new platforms for making contact with social groups, such as researchers' radio programs on local stations or documentaries, can be observed here.

While scholars near the borders have appropriated the migration issue as a situation in which they are embedded, those in the centre approach the issue more "scientifically." Detached from the daily concerns of the border, these scholars are less keen on Open Science (i.e., Open Research Agenda or Open Data), and migration as a public issue is co-produced in relation to mainstream academic literature and, indirectly, to the federal government (e.g., consultancy and advisory). Thus, the emergence of public problems is not only conflictive but also dynamic because it is the outcome of a process of interacting actors in different places.

Conclusions

Over the course of this chapter, we have examined the conditions under which scientific knowledge (produced in a more or less open way according to each particular case) is capable of being utilized to satisfy social needs. Our approach takes the social use of knowledge as its focus, not "subsequent" to its production but co-produced with it. In this way, we are inserted into a concrete dynamic of elaboration in conjunction with closure/openness of scientific knowledge. At the same time, this perspective enables us to glimpse the given (and changing) forms or conditions of relationships in which these dynamics acquire a certain entity. Thus, from an analysis of the cases presented, some meaningful dimensions about openness emerge which help us to make advances on our area of study. What is valuable about these dimensions is that they show the concrete framework in which human activity unfolds, accounting for vital aspects which, up until now, have been scarcely and superficially tackled in the mainstream of Open Science: the competencies, skills, organizational forms, and social resources (economic, political, and cognitive) deployed by the actors constituted in the knowledge productive processes.

We confront different configurations of public problems/issues as social and cognitive realms, which delimit that which is disputable, expressible, and cognizable. This is a nodal aspect to the question of the relationship between openness and the utilization of knowledge, given that the definition of the problematic framework makes certain knowledge possible, which far from being purely universal is utilizable in the realm of certain relationships and by certain actors within them. But these problems, far from being "natural," are actively constructed by the actors, mobilizing diverse types of knowledge.

The role of a certain type of actor, that of the "driver," is strikingly significant: an actor who in some way marks an initiation or rupture, mobilizing scientific knowledge in a particular way in pursuit of a particular social use, and is a highly active and influential element in shaping the public problem. Furthermore, the constitution of the public problem can be characterized by varying degrees of conflict. The degree of conflict participates in its configuration as well as in the possibilities of intervention available to other actors, who mobilize their own resources, organizational forms, competencies, and skills, and give rise to the configuration of new types of knowledge. This requires, however, the possession of specific competencies and resources by the affected actors, as well as certain forms of production, mediation, intermediation, and stabilization of the knowledge in question.

On the other hand, we regard the material dimension of knowledge to be significant in relation to the possible forms of "use" and "openness." This dimension does not determine the practices of production and use of knowledge, but it does facilitate certain "conditions of possibility" for the establishment of more or less collaborative relations of production, access to the products of science, and their eventual (re)use. The material dimension is definitively linked to the other dimensions of co-production, and they are therefore able to mutually modify themselves (and each other) according to different contexts. In the case of Chagas disease, molecular biologists have typically imposed their own perspective on the public problem and function as "drivers" of the process. The way in which knowledge is made utilizable is in the form of scientific publications, or rather as information outputs codified and organized into databases (DNA sequences). This form entails certain qualifications that would allow one to mobilize and use these resources. Therefore, to facilitate uses of knowledge that would be commensurable with social needs and demands, very specific processes of translation are required to convert them into commercializable pharmacological products or new therapeutic devices. These processes involve, in turn, another realm of relations, actors, resources, competencies, and organizational forms, as well as a different overall relation with the object of research.

The question is quite different for the technical reports in the Jáchal-Veladero case. In that case, the translation does not operate upon the material form of the knowledge but on its socio-cognitive content and, likewise, on its problematic criteria of elaboration. This conversion implicated particular forms of social relations, characterized by modalities of collaborative knowledge production, certain modes of political organization—such as the constitution of AJNST itself and the web of political alliances with diverse groups—and, lastly, certain types of socio-cognitive skills such as reasoning about diverse technical problems, defining analysis criteria, drafting reports, mobilizing university laboratories, and selecting and enlisting national and international scientific actors.

On the other hand, it can be observed that different disciplinary regimes have a significant influence on the intersections between "use of knowledge" and "Open Science." In principle, the regimes anchored to a single, strongly established discipline integrated into international agendas seem to be guided more by the legitimization of knowledge through the classic means of circulation (articles in high-impact journals) than by their relationships with an approach to public problems, even when public discourse seems to be contradictory. This is the case with molecular biology and applied genomics in the study of *T. cruzi*, in which despite the formally "open" character of knowledge, a set of specific competencies is required for access. These competencies operate as serious "barriers to entry," both for the "non-specialist" scientists (or those in peripheral contexts) and, in the same sense, for the industrialization of knowledge, which could be appropriately used in the previously defined social problem. The participation of "non-scientist" actors is, here, highly limited.

In contrast, the processes of co-production around the Jáchal socio-technical dispute unfolded through the confluence of various disciplinary fields with a technical character and a lesser degree of specialization and international integration, which contributed to producing a scenario characterized by lower levels of restriction. Thus, conditions arose that enabled the AJNST, constituted by a non-scientific public and its "non-specialist" scientist allies, to intervene with remarkable depth and impact.

In the case of strategies for conserving threatened species, although the "driver" was originally situated within the field of environmental studies or ecological conversation. This field is, in itself, less structured along disciplinary lines than molecular biology and, in its own process of development, is more permeable than molecular biology. On the other hand, the knowledge mobilized is gathered from various prior objects constituted as research problems: soils, environmental systems, climatic systems, studies of human action, etc. The cognitive problems of this type of discipline, whose empirical objects are distributed on a wide-ranging regional scale (as is the case with conservation biology), mean a greater possibility of openness toward the process of open collaboration, both in terms of the use of technological infrastructures and the intensive use of human collaborators.

The social sciences, in this disciplinary regime, present special features. On the one hand, they make the boundaries between knowledge producers and the data-providing subjects more nebulous; on the other hand, the frontiers between the different disciplines (anthropology, history, sociology) are less clearly demarcated, unlike, for example, approaches in the "hard" sciences. The disciplinary investigations are inscribed into paradigms as diverse as the more "scientific" research (more distanced from the subjects) that only permits access to data once they have been crystallized as such to that of "action research," which is much closer to the notion of "science-social actor" co-production and in which the use of knowledge is constitutive of said epistemic activities.

We observed that in the case of Jáchal, the socio-technical dispute is inscribed in a context of productive relations that give rise to "exclusive knowledge," since the "opencast" mine barely provides work or resources for the local populations and is, furthermore, incompatible with the technological forms that are effectively utilizable in the pursuit of meeting social needs. As we have seen, this type of exclusive knowledge is opposed by a type of knowledge mobilized by other "drivers" who question the public definition of the problem as well as the closed character of the knowledge mobilized by the company and the actors associated with it.

In the case of the participatory strategies in environmental conservation in which the configuration of actors starts from a "driver" who distributes information-collecting tasks between diverse actors, the participation of citizens as information gatherers entails a degree of instrumentalization of the process of openness, while the processing and analysis of the data are left to the experts.

In the research looking at social sciences in Mexico, the drivers, evidently, are the social scientists. But here, unlike the other cases, the frontiers become more diffuse since the knowledge is produced—as in much research in the social sciences—by the researchers interacting with the studied subjects (or communities). Therefore, even if the social subjects are held to be "mere providers of information," they, necessarily, have their own representations about the problem (i.e., of migration) and a de facto tension arises between their own interpretations and those of the scientists. Varied degrees of integration are therefore possible in this context, and the uses of knowledge obtained can be the object of disputes, with greater or lesser remoteness from the cognitive dimensions.

The concepts and cases discussed thus far could help us articulate explicit recommendations for enabling more effective uses of scientific knowledge on behalf of local stakeholders. Public problems are processes whereby unequally distributed resources become mobilized. Therefore, the affected groups could take advantage of spaces where their position is strengthened. These spaces become even more crucial in the Latin American context, where the number of well-established or institutionalized spaces that allow knowledge to circulate openly are scarce. A diversity of stakeholders and modes of approaching public problems and intervention should be required to integrate these spaces.

Strengthening and institutionalizing public science forums could be a way to foster the mobilization and production of knowledge aimed toward addressing social needs and demands. The challenge lies in ensuring legal state support while at the same time enabling local stakeholders to retain their autonomy against potential mechanisms of co-optation induced by political, scientific, or economic corporatism.

Public science forums could contribute to scientific openness in the usual sense, but they also may allow alternative forms of knowledge born by different stakeholders—usually deemed as inferior or "non-scientific"—to take part in public debate and intervention. The affected stakeholders and their own sets of knowledge could therefore participate in both the formulation and the resolution of the problems at stake. Public science forums can also affect public deliberation by providing policy-making with different grounds. This is clearly crucial in the process of intervening in public controversies and their outcomes, but also in non-controversial issues where more reflexive and representative criteria are needed to ensure that knowledge will be effectively used, such as in the cases of Chagas disease research and wildlife conservation.

These kinds of forums may also enable the application of the precautionary principle, which in recent times has been particularly difficult in Latin America, especially in the face of environmental hazards. As we have shown through the mining controversies in Jáchal, there are cases where conflicts are settled through the interposition of technical reports. Lay groups, in general, and potentially affected groups, in particular, are usually sidelined from the elaboration of technical reports and their consequent decision-making processes. Public science forums aim to revert this power imbalance in both political and cognitive terms. In this way, our proposal is partially amenable to the ideas of citizen science, in the sense that it calls for a systematic fostering of institutional spaces for both scientific openness and political participation. However, it also goes beyond citizen science as it understands scientific knowledge and public processes as co-producing each other, rather than just knowledge outputs that "inform" political decision-making.

Notes

1. The disaggregation of this goal took the form of a series of research questions: Where are the jaguars (and pumas) found in the Atlantic forests? What features must the "landscape" possess for the species to subsist (D'Angelo 2009)? What factors determine population density variation? How many jaguars are there in the region (Paviolo 2010)?

References

Agüero, Fernàn, Bissan Al-Lazikani, Martin Aslett, M. Berriman, Frederick S. Buckner, Robert K. Campbell, and Christophe L. M. J. Verlinde et al. 2008. "Genomic-scale Prioritization of Drug Targets: the TDR Targets Database." *Nature Reviews: Drug Discovery* 7 (11): 900–7. http://doi.org/10.1038/nrd2684.

Alatas, Syed F. 2003. "Academic Dependency and the Global Division of Labour in the Social Sciences." *Current Sociology* 51 (6): 599–613.

Arocena, Rodrigo, and Judith Sutz. 2001. "Changing Knowledge Production and Latin American Universities." *Research Policy* (30) 8: 1221–34.

Behague, Dominique, Charlotte Tawiah, Mikey Rosato, Télésphore Some, and Joanna Morrison. 2009. "Evidence-based Policy-making: The Implications of Globally-applicable Research for Context-specific Problem-solving in Developing Countries." *Social Science and Medicine* 69 (10): 1539–46. http://doi.org/10.1016/j.socscimed.2009.08.006.

Calvillo, Arturo. 2015. "Migración centro americana en México, invisible para el gobierno." *HipanTVNexo Latino*. www.hispantv.com/noticias/reporteros /75490/migrantes-centroamerica-violencia-sexual-mexico-eeuu.

Comisión Interamericana de Derechos Humanos (CIDH). 2016. *Situación de Derechos Humanos en México*. www.oas.org/es/cidh/informes/pdfs /Mexico2016-es.pdf.

Crawshaw Peter, G., Jr. 1995. "Comparative Ecology of Ocelot (Felis pardalis) and Jaguar (Panthera onca) in a Protected Subtropical Forest in Brazil and Argentina." PhD Thesis. Gainesville: University of Florida.

De Angelo, Carlos, Agustín Paviolo, Daniela Rode, Laury Cullen et al. 2011. "Participatory Networks for Large-scale Monitoring of Large Carnivores: Pumas and Jaguars of the Upper Parana Atlantic Forest." *Oryx* 45: 534–45.

De Ángelo, Carlos. 2009. "El paisaje del Bosque Atlántico del Alto Paraná y sus efectos sobre la distribución y estructura poblacional del jaguar (Panthera onca) y el puma (Puma concolor)." PhD Dissertation. Buenos Aires: Universidad de Buenos Aires.

Díaz, Elena, Yolanda Texera, and Hebe Vessuri. 1983. *La ciencia periférica*. Caracas: Monte Ávila Editores.

Di Bitetti, Mario S. 2015. "Cómo estudiar poblaciones de mamíferos silvestres." *Revista Ciencia Hoy* 25 (146): 31–7.

Etzkowitz, Henry, and Loet Leydesdorff. 2000. "The Dynamics of Innovation: From National Systems and 'Mode 2' to a Triple Helix of University-Industry-Government Relations." *Research Policy* 29 (2): 109–123.

Fecher, Benedikt, and Sascha Friesike. 2014. "Open Science: One Term, Five Schools of Thought." In *Opening Science*, edited by Sönke Bartling and Sascha Friesike, 213–24. Springer International Publishing. https://doi .org/10.1007/978-3-319-00026-8.

Gusfield, Joseph R. 1981. *The Culture of Public Problems: Drinking-driving and the Symbolic Order*. Chicago: The University of Chicago Press.

Inter-American Development Bank (IADB). 2001. *Competitiveness: The Business of Growth*. Washington, DC: IADB.

Hotez, Peter J., Eric Dumonteil, Miguel Betancourt Cravioto, Maria Elena Bottazzi, Roberto Tapia-Conyer, Sheba Meymandi, Unni Karunakara, Isabela Ribeiro, Rachel M. Cohen, and Bernard Pecoul. 2013. "An Unfolding Tragedy of Chagas Disease in North America." *PLoS Neglected Tropical Diseases* 7 (10). http://doi.org/10.1371/journal.pntd.0002300.

Jasanoff, Sheila. 2004. "The Idiom of Co-production." In *States of Knowledge: The Co-production of Science and Social Order*, edited by S. Jasanoff. London: Routledge.

Kreimer, Pablo. 2003. "Conocimiento científicos y utilidad social." *Ciencia, Docencia y Tecnología*, AñoXIV, 26.

————. 2014. "'Citizen of the World' or a Local Producer of Useful Knowledge? That's the Question." In *Academic Dependency and Professionalization in the South*, edited by Fernanda Beigel and Hanan Sabea.

————. 2015. "Co-producing Social Problems and Scientific Knowledge: Chagas Disease and the Dynamics of Research Fields in Latin America." In *Sociology of Science Yearbook*. Springer.

Kreimer, Pablo, and Thomas, H. 2005. "What is CANA-AKNA? Social Utility of Scientific and Technological Knowledge: Challenges for Latin American Research Centers." In *Development Through Knowledge*, edited by Jean-Baptisete Myer and Michel Carton.

Kreimer, Pablo, and Juan Pablo Zabala. 2007. "Chagas Disease in Argentina: Reciprocal Construction of Social and Scientific Problems." *Science Technology & Society* 12 (1): 49–72. http://doi.org/10.1177/097172180601200104.

Kreimer, Pablo, and L. Levin. 2013. "Scientific Cooperation between the European Union and Latin American Countries: Framework Programmes 6 and 7." In *Research Collaborations between Europe and Latin America: Mapping and Understanding Partnership*, edited by Gaillard and Arvanitis. Paris: Editions des Archives Contemporaines.

Leys Stepan, Nancy. 2011. *Eradication: Ridding the World of Diseases Forever?* Ithaca: Cornell University Press.

Losego, Philippe, and Rigas Arvanitis. 2008. "La ciencia en los países no hegemónicos," *Revue d'anthropologie des connaissances* 2 (3): 351–9. http://doi.org/10.3917/rac.005.0351.

Lundvall, Bengt-Âke. 1992. National Systems of Innovation: Towards a Theory of Innovation and Interactive Learning. London: Pinter.

Magarinos, María P., Santiago J. Carmona, Gregory J. Crowther, Stuart A. Ralph, David S. Roos, Dhanasekaran Shanmugam, Wesley C. Van Voorhis, and Fernán Aguero. 2012. "TDR Targets: A Chemogenomics Resource for Neglected Diseases." *Nucleic Acids Research* 40 (D1): D1118–27. http://doi.org/10.1093/nar/gkr1053.

Masum, Hassan, and Rachelle Harris. 2011. *Open Source for Neglected Diseases: Magic Bullet or Mirage?* Washington, DC: Results for Development Institute.

OCDE. 1996. "La difusión de tecnología," *Redes: Revista de Estudios Sociales de la Ciencia* III (8).

Ojeda, Ricardo, V. Chillo, and Gabriela Diaz Isenrath. 2012. *Libro Rojo de Mamíferos Amenazados en Argentina*. Mendoza: SAREM.

Paviolo, Augustín, Carlos Daniel De Angelo, Yamil E. Di Blanco, and Mario S. Di Bitetti. 2008. "Jaguar Pantheraonca Population Decline in the Upper Paraná Atlantic Forest of Argentina and Brazil." *Oryx* 42 (4): 554–61.

Paviolo, Augustín. 2010. "Densidad del yaguareté (Panthera onca) en la selva paranaense: su relación con la disponibilidad de presas, presión de caza y coexistencia con el puma (Puma concolor)." Tesis de Doctorado. Cordoba: Universidad Nacional de Córdoba.

Porrás, Analía I., Zaida E. Yadon, Jaime Altcheh, Constança Britto, Gabriela C. Chaves, Laurence Flevaud, et al. 2015. "Target Product Profile (TPP) for Chagas Disease Point-of-Care Diagnosis and Assessment of Response to Treatment." *PLOS Neglected Tropical Diseases* 9 (6). http://doi.org/10.1371/journal.pntd.0003697.

Sábato, J., and N. Botana. 1969. "La ciencia y la tecnología en el desarrollo de América Latina," in Herrera, Amilcar O. (ed.), *América Latina: ciencia y tecnología en el desarrollo de la sociedad*, Santiago de Chile: Editorial Universitaria.

Sosa-Estani, S. 2011. "Nuevo escenario de estudios clínicos y perspectivas para los próximos años." *Plataforma de Investigación Clínica En Enfermedad de Chagas* 11.

Sutz, Judith. 1994. *Universidad y Sectores Productivos*. Buenos Aires, Centro Editor de América Latina.

Whitley, Richard. 2012. "La organización intelectual y social de las ciencias." Universidad Nacional de Quilmes.

WHO. 2007. "New Online Database to Help Fight Infectious Diseases." http://www.who.int/mediacentre/news/releases/2007/pr14/en/#.U1gFkMar3Vs.mendeley.

WHO. 2012. "Research Priorities for Chagas Disease, Human African Trypanosomiasis, and Leishmaniasis." Geneva, Switzerland Geneva: WHO.

CHAPTER 10

Tensions Related to Openness in Researching Indigenous Peoples' Knowledge Systems and Intellectual Property Rights

Cath Traynor, Laura Foster, and Tobias Schonwetter

Abstract

This chapter explores issues of boundaries in practices of Open Science regarding research involving Indigenous peoples in South Africa. We start considering colonial notions of "science" and "openness," and how historical injustices and lack of redress influence the context in which our current research sits. Our research broadly aimed to develop a political, ecological approach to understanding the relationship between climate change, intellectual property, and indigenous peoples. Our approach was influenced by "decolonizing methodologies" and feminist perspectives, and we employed participatory action research methodologies to guide not just the substantive, but also procedural elements of the research. We discuss our experience with developing "community-researcher contracts" in an attempt to make ourselves as researchers more accountable to Indigenous Nama and Griqua communities and to adequately protect their Indigenous knowledge. The challenges of negotiating the contracts is described and how we conceptualized the concept of a "situated openness"—a way of doing research that assumes knowledge production and dissemination is situated within particular historical, political, socio-cultural, and legal relations.

Introduction

This chapter offers preliminary field notes on the practice of engaging in an open and collaborative research project that involves multiple scales of histories, geographies, institutions, and ways of knowing. We have been engaged in a two-year collaborative project with a team of Indigenous community leaders, academics, and lawyers examining issues of climate change and Indigenous knowledge. Our team includes Cecil Le Fleur (Griqua National Council) and Gert Links (Richtersveld Traditional Nama Council), as well as the three authors of this chapter, Laura Foster (Indiana University), Tobias Schonwetter (University of Cape Town), and Cath Traynor (Natural Justice). Our project was supported by the Indigenous leaders, who suggested we interview members of their communities concerning their understanding of the impact of climate change and the role of Indigenous knowledge in climate change adaptation.

In doing this research, we have been mindful of how the varied geographies of Nama, Griqua, South African, and American nations and the multiple histories of colonialism, apartheid, and post-apartheid shape our work. We have learned much from navigating the different institutional worlds of Nama and Griqua councils, academia, and non-profits. We have also become more familiar with the different and similar traditions of producing and disseminating knowledge that each of us are located within—Indigenous peoples' knowledge, feminist studies, scientific studies, legal studies, and ecology. In doing this work, we have focused on the very process of doing research in order to understand the relations of power that enable and limit possibilities for open and collaborative research. A central finding of our research has been that efforts to adapt to climate change, which involve or will impact Indigenous peoples or their lands and resources, must begin with developing more socially just ways of doing research.

As principal investigator and manager of this collaborative project, Cath Traynor's learning and contributions to the project were guided by her experience as an ecologist and non-profit practitioner with Natural Justice. The mission of Natural Justice is to work collaboratively with those Indigenous and local communities who seek them out for legal expertise on how to secure their rights to land, resources, knowledge, political representation, and self-determination more broadly. Her main interest in the project was therefore aimed more at developing practical strategies for adequately protecting

Indigenous peoples' knowledge. What became apparent, however, was the need for community-based research contracts that would ensure practices of open and collaborative research that meet the needs and interests of Indigenous peoples. Although the drafting and negotiating of these contracts are ongoing, we offer some initial fieldnotes here on how certain policies at the international, national, and university level shaped the drafting of these contracts and the collaborative research practices that they seek to promote. Open and collaborative research requires recognition of different knowledge and writing practices; thus, our choice to frame this chapter as fieldnotes is a deliberate attempt to push back upon the hegemony of academic scholarly expectations that can hinder truly meaningful collaborative research practices.

Historical Background and Conceptual Framing of the Project

Nowadays, open and accessible systems and practices are seen in many areas as a crucial engine for innovation and socio-economic development, particularly in Africa through, among other things, facilitating collaboration and improving transparency and accountability. But openness is not an end in itself, and there are potential downsides to openness, especially if only some elements of openness are asserted in a one-sided, exploitative, and selective fashion. Where this has happened, a more nuanced, "situated" approach to openness is required to account for past injustices and to prevent further harm to those affected. This collaborative project requires a situated approach to openness as the notions of science as "open" and nature as "freely accessible" have historically been invoked to exploit countries such as South Africa. For example, British and Dutch colonial scientists characterized resources in South Africa as "belonging to no one" under the doctrine of *terra nullius* in order to take biodiverse plants and produce botanical science. To the extent that their activities involved appropriation of such materials and research results, the colonial scientists appeared, however, to be less concerned about openness and free accessibility for all. Indeed, the *terra nullius* doctrine was not restricted to science, but widespread among colonial authorities, who used the principle and that of *mise en valeur*[1] to justify land seizures from Indigenous peoples, most of whom were mobile land users such as hunter gatherers or pastoralists and therefore did not meet the colonists' criteria for occupation of lands (UN 2012).

In these cases, the notion that knowledge and resources should be open and accessible has therefore been historically misused to cast countries in the Global South, including South Africa, as suppliers rather than producers of knowledge, and in particular, Indigenous peoples' knowledge, resources, and heritage as free for the taking. Furthermore, Indigenous communities within South Africa continue to face historical injustices as colonization, apartheid, and post-apartheid laws and policies have not fully taken their unique ways of life and culture into account. Thus, similar to other Indigenous communities in Africa, they have been severely marginalized, and many rights and freedoms enjoyed by their fellow citizens are inaccessible to them (Barume 2010). Most recently since the signing of the Constitution of South Africa in 1994, Indigenous communities have been engaged in struggles to right previous wrongs. For example, Nama pastoralists in Richtersveld initiated a court case[2] in South Africa to reclaim the tenure of their ancestral lands. The Constitutional Court of South Africa ruled in favour of their land and mineral rights; as a result, in 2002, the International Criminal Court set aside[3] the use of *terra nullius* as a justification for disenfranchisement. Presently, Indigenous representatives continue to advocate for recognition of Khoi and San customary governance structures; indeed, the Traditional and Khoi-San Leadership Bill (TKLB) has been introduced to the National Assembly, one of its key objectives being to recognize Khoi and San leaders in the formal, traditional leadership structures of South Africa. However, the Bill also seeks to address additional issues related to other traditional communities; thus, the Bill is highly contested by many communities, academics, and civil society (e.g. Makoena 2015).

The broad objective of our project was to develop a political, ecological approach to understanding the relationship between climate change, intellectual property, and Indigenous peoples. This approach sought to understand the relationship between these three facets and how political, economic, legal, historical, and socio-cultural processes structure them. The project employed participatory action research (PAR) design and methods with the aim of reducing the power relations within and between researchers/researched and hierarchies of knowledge production by involving marginalized groups within the design, implementation, and outcomes of the research. Rather than studying communities from the "top-down," PAR takes a "bottom-up" approach by forming partnerships with communities to identify key issues of importance and develop ways of doing research, interpreting results, and taking action on the findings (Smith et al. 2010). This enables the research to better respond to the interests and needs of the community in ways that benefit them (Maguire 1996).

Our approach was informed by Linda Tuhiwai Smith's (1999) concept of "decolonizing methodologies" that demonstrates how research practices have historically contributed to the colonization of Indigenous peoples. Models of Western knowledge production have been positioned as superior, which has engendered the devaluing of Indigenous peoples' knowledge. Furthermore, we were also cognizant that the institutions we as researchers are part of (universities and an NGO) can be colonizing spaces themselves, and that we should also engage mindfully with the research requirements and processes of our own institutions. Our aim was to "decolonize" historical modes of producing knowledge by positioning Indigenous peoples as producers of climate change knowledge through open and collaborative PAR processes.

Given the histories noted above, our project was guided by an understanding of "situated openness." Appeals for open and collaborative research are often based upon understandings of an open public domain where data and research results are meant to be freely shared and open to others. In arguing for a situated public domain, Laura Foster contends that norms of openness and sharing have historically been deployed by researchers to appropriate and exploit Indigenous peoples' lands, knowledge, and resources (Foster 2011). A situated public domain is alternatively based upon norms of openness and protectiveness that allow Indigenous peoples to decide for themselves when, how, and to what extent their knowledge should be shared (Foster 2011). Building upon these insights, our project is framed through an understanding of a situated public domain that also demands a model of situated openness. Drawing upon Foster's work and feminist science studies broadly, the understanding of situated openness requires us to consider how collaborative knowledge production is situated within particular historical, political, sociocultural, and legal relations of inequality. Collaborative knowledge practices based upon norms of openness can democratize knowledge, but can, as mentioned above, also be misused to legitimize the taking of Indigenous peoples' knowledge. What is needed are practices of collaborative knowledge production that involve simultaneous modes of being open, closed, sharing, and restrictive in order to democratize science in more meaningful ways for Indigenous peoples.

Community-Researcher Contracts

The demands for data and research results to be open and accessible to others created some tensions with our desire to protect Indigenous peoples' knowledge and knowledge holders' interests. Indigenous peoples have experienced histories of violence that have led to the taking of their lands, knowledge, and heritage—this includes experiences with academic researchers, even up to the present day. As a result, Indigenous peoples are sometimes less willing to share their knowledge freely without prior informed consent and meaningful collaborative consultation.

To counter those histories, our team sought to develop 'community-researcher contracts' between Natural Justice, Indiana University, the University of Cape Town, and the Nama and Griqua communities.[4] These contracts are meant to clearly state expectations and responsibilities between parties, how the research will be conducted, and how knowledge may (or may not) be shared.

We also wanted to ensure that Indigenous knowledge (IK) and knowledge holders' rights were protected in line with international laws. Several international law instruments specifically refer to IK[5]: for example, the UN Declaration on the Rights of Indigenous Peoples (UNDRIP) states that Indigenous peoples have the right to maintain, control, protect, and develop their traditional knowledge and the manifestations of their science (UN 2008, 11). The UN Convention on Biological Diversity (CBD) states that each Contracting Party to the Convention shall respect, preserve, and maintain the knowledge, innovations, and practices of Indigenous peoples (UNEP 1992, 6). However, there is no universally agreed-upon definition of IK, it is not addressed uniformly by the different instruments, and some of these instruments seek to protect IK by restricting access and use (Savaresi 2016). Furthermore, interpreting how these instruments and processes impact IK–related research in the relatively new field of climate change requires expert guidance. Indeed, many so-called "soft" international instruments such as the Nagoya Protocol of the CBD and the UNFCCC Paris Agreement give deference to national laws; thus, an understanding of the national legal landscape regarding IK is essential (Savaresi 2016).

At the national level, there may be specific laws and policies that simultaneously recognize international rights related to IK but also undermine them. Our project was focused on South Africa, which, since the end of formal apartheid rule in 1994, has been developing new laws and policies related to indigenous knowledge systems (IKS). Currently, the pivotal policy is the Indigenous Knowledge Systems Policy that is designed as an enabling framework to stimulate and strengthen the contribution of IK to social and economic development in South Africa (Republic of South Africa 2004). One of the key policy drivers is the affirmation of African cultural values in order to redress histories of subordination under apartheid rule whereby IKS and its practitioners were marginalized, suppressed, and subjected to ridicule. Furthermore, this policy notes that in regard to the protection of IKS, South Africa has a well-defined system of intellectual property rights; however, legal strategies for the perpetual protection of IKS through benefit sharing and/or joint ownership are continuing to be debated. To address shortfalls, a Draft Protection, Promotion, Development and Management of Indigenous Knowledge Systems Bill (hereinafter IKS Bill) was introduced into Parliament in 2015[6] and amended in 2016.[7] The intention of this IKS Bill is laudable as it aims to establish a unique, so-called *sui generis* approach for the protection of IK instead of relying on existing IP frameworks to provide for such protections (Schonwetter, Jansen, and Foster 2015). The IKS Bill states that the ownership of Indigenous knowledge vests in the Indigenous community, that a trustee of the Indigenous community can hold the IK in trust on behalf of the community, and that this trustee shall be responsible to the community for the protection of their rights (Republic of South Africa 2016). However, as the IKS Bill is still under discussion and may change considerably, Natural Justice, as legal advisors to the communities, engaged by taking expert legal guidance on specific issues that would be in the best interests of the IK knowledge holders and discussing the various options and implications of specific text with community representatives. In addition to the IKS Bill, South Africa recently issued a draft Indigenous Knowledge Systems Research Ethics Policy[8] that aims to protect communities and their IK, reduce the adverse effects of research, ensure that communities equally own data and information generated by the research, and ensure fair and equitable benefit sharing arising from the communities' contributions to the research process. The Research Ethics Policy also emphasizes full informed consent and principles of confidentiality, empowerment, and prior rights.[9]

Difficulties Developing the Community-Researcher Contract

In developing the specific text of the "community-researcher contracts," our project team experienced several difficulties, some of which are elaborated in more detail below. Because the project was a collaboration with universities as outlined above, we gained approval prior to starting research from the UCT research ethics committee (REC) and IU institutional review board (IRB). The REC/IRB approval certainly helped to ensure ethical conduct of human subjects research; however, we found that it very much focused on the individual and assumes that knowledge is individually held, which was incongruent with how indigenous Nama and Griqua communities related to their knowledge.[10] These communities hold their knowledge collectively; thus we reasoned, should we also obtain collective consent from the community prior to conducting research and sharing our research outputs? In terms of the research process and timelines, we faced a conundrum. Although we could obtain collective buy-in from the leaders of the community prior to conducting research, the exact nature of the knowledge shared would not be known, which made obtaining collective community consent difficult. We were committed to obtaining individual consent from individuals with whom we spoke. We were also committed to returning to the community and/ or their representative leaders to share with them what we learned and seek their collective consent to use and share our learning in our research.

The Community Research Contract needed to elaborate this dynamic process to ensure the collective element of IK was addressed and included as part of a broader ethics clearance process. It also needed to address several concerns: If Nama and Griqua peoples shared Indigenous knowledge with us, what safeguards were needed to avoid misappropriation? If our funders required us to make our "data" open and freely accessible, how could we fulfill these funder requests while ensuring adequate protection of indigenous Nama and Griqua communities? Indigenous peoples' knowledge must not be publicly disseminated without their free, prior, informed consent (FPIC) at each stage of the research and its dissemination.

Communities have a right to FPIC, and an important part of this is "complete disclosure of the risks and benefits to individuals and the community of participation in the research" (Republic of South Africa n.d.). Thinking through different scenarios that could arise from sharing different elements of IK is required so these are considered upfront. For example, pastoralists may share the specific characteristics of their breeds of livestock, which enable the animals to cope with harsh environmental conditions such as excessive heat, drought, and limited forage. These characteristics could potentially be very valuable to other livestock keepers and breeders (including commercial breeders), and inadequate protection could increase the risk of misappropriation for the community.

As for international and domestic legal frameworks, how one interprets certain provisions is often key, and we found expert guidance from lawyers with practical experience in supporting Khoi and San communities in Southern Africa when negotiating access and benefit-sharing agreements regarding their IK valuable. Additionally, we needed to continually ask ourselves, "What does this mean in practice?" Interpreting the meaning of legal texts was no easy matter, and developing clear, practical statements and actions for the community-researcher contracts to ensure adherence was challenging and, at times, overwhelming.

Our project is an international collaboration; from a legal perspective, we also needed to consider that different laws and policies apply in different countries. For instance, South Africa's laws and policies are, of course, only applicable within the country's geographic boundaries, and South Africa is only bound to the international legal instruments to which it has adhered. We also needed to consider foreign legislative frameworks to determine what happens to the IK and knowledge holders' rights when the IK leaves South Africa. This was particularly pertinent in our case, as one of our partners was based in the United States, which is a country that has signed but not ratified the CBD.[11] Thus, protections such as those offered under the Nagoya Protocol on Access and Benefit Sharing would not be fully available to govern our research.[12] The contract therefore needed to address such gaps.

In addition to harnessing laws and policies that could support the communities and their IK, we also needed to examine policies that may undermine knowledge holders' rights. For example, intellectual property law may work to undermine their rights due to the stark differences between the nature of knowledge, property, and ownership in Western law and Indigenous customary laws and worldviews (Natural Justice 2015). South Africa's IKS Bill is commendable as it aims to establish a *sui generis*[13] or an intellectual approach to the protection of IK, which would then provide indigenous communities with different options to protect and manage their IKS (Schonwetter, Jansen, and Foster 2015). However, the Bill is, as mentioned above, still in draft form and thus the *sui generis* option is not currently available. Due to these gaps in protection for IK, we sought to develop a community research contract that would provide adequate protection. The community research contract also sought to establish protections for Indigenous peoples as specified in South Africa's Draft IKS Research Ethics Policy (Republic of South Africa n.d.).

The contracts are to be concluded between the Indigenous groups and the universities themselves, so they hold the institutions more accountable. The very process of negotiating these contracts has increased research communications between parties and has revealed how university policies and procedures can prevent practices of collaborative science. For example, a key purpose of the Draft IKS Research Ethics Policy is "to ensure...that the communities equally own data and information generated or produced." However, one university objected to joint-ownership because of a lack of clarity regarding who controls decisions over what is done with the research materials; thus, creating joint ownership can be problematic in practice.

Developing the contracts has involved a series of back-and-forth discussions and negotiations over specific contractual provisions. The contract, for example, now specifies that researchers must agree not to share Indigenous peoples' knowledge without their consent, to respect Indigenous peoples' intellectual property rights, and to not produce knowledge that would harm the reputation of the community. In negotiating these contractual provisions, we have begun to identify the precise university policies and procedures that hinder collaborative research practices with indigenous Nama and Griqua peoples. We have also begun to understand the limitations of community research contracts. Although the contractual provisions are meant to disrupt hierarchies between researchers and researched, it is unclear if contracts are the appropriate vehicle for reducing hierarchies of knowledge production. Only those who sign the contracts are bound by them for the specified duration, which means that third parties having access to the IK are not bound by the responsibilities set out in the contracts.

Conclusion

The field notes shared in this chapter have reflected on tensions related to openness in research with Indigenous peoples on issues related to their knowledge systems and intellectual property rights. Although the detailed findings are specific to our particular case, they provide insights highly relevant for practitioners of open and collaborative science working together with historically marginalized groups, such as Indigenous peoples.

Our example illustrates the importance of considering contexts in which the current research is situated, and that Open Science practitioners need to acknowledge injustices faced by Indigenous communities both historically and in the present day. Researchers, together with communities, need to strive to develop research methodologies and processes that speak to the need for redress. Our experiences show that simply meeting the ethical research requirements of academic institutions is not enough; researchers need to critically engage with these structures, identify where they fall short, and then find creative ways to address the gaps. Ethics approval processes that are based upon the notion that knowledge is individually held will not meet the needs of many Indigenous communities who view their knowledge as being collectively held.

Open Science practitioners need to consider legal protections for Indigenous knowledge prior to sharing. Although there are some positive protections available under international instruments such as the Nagoya Protocol, these have limitations. Understanding national protections for IK and what they mean in practice is key. Our South African case study illustrates the dynamism of the legal system, and although a unique *sui generis* system is under development in the IKS Bill, it is not yet available. Additionally, at the national level, existing intellectual property laws can undermine IK as they do not meet its needs. Thus, prior to sharing IK, legal insufficiencies need to be addressed.

We employed contracts as a tool to address limitations within institutional ethics processes and international and national laws. Developing and negotiating these has led to positive results, such as increased communication between parties and deeper understanding of critical issues with regard to IK protection. However, by working through what contracts mean in practice, we have also identified several potential barriers related to the mission and policies of academic institutions, which could prevent truly collaborative science processes and also limit protection for communities and their IK. Contracts certainly have the potential to address some shortfalls in existing research processes, but they are no panacea. Thus, when engaging in Open Science practices with researchers from Indigenous communities, their institutions, and funders must acknowledge there will be certain boundaries to openness and be cognizant of situated openness models. Furthermore, Indigenous communities must be fully informed and legally empowered to negotiate their own terms relating to research processes so they meet their unique needs.

Notes

1. The colonial discriminatory concept that only cultivation of land by crop production was an effective use of land.
2. The Alexor Ltd and another vs. Richtersveld Community and Others case.
3. Meaning the term has no standing and its legal authority is removed.
4. Natural Justice as lawyers and the Project Manager for the research led this process with the aim being to ensure protection of the communities and their IK. Natural Justice liaised with the Traditional Leaders, government, and the university's legal/faculty representatives to develop these contracts. The academic researchers stepped back from this process; as university employees, it was a possible conflict of interest for them to promote the communities' needs above the research needs of their respective universities.
5. Including the Convention on Biological Diversity (Rio de Janeiro, 5 June 1992, in force 29 December 1993), UN Convention to Combat Desertification in Countries Experiencing Serious Drought and/or Desertification, Particularly in Africa (Paris, 14 October 1994, in force 26 December 1996), UN Framework Convention on Climate Change (New York, 9 May 1992, in force 21 March 1994)—Paris Agreement (Paris, 12 December 2015, ratification in process, not yet entered into force as at 20.09/2016).
6. Draft Protection, Promotion, Development and Management of Indigenous Knowledge Systems Bill 2014, General Notice 243 of 2015 (GG 38574, 20 March 2015).
7. Protection, Promotion, Development and Management of Indigenous Knowledge Systems Bill 2015 (amended) Notice of Introduction of a Bill into Parliament, Notice 199 of 2016, Department of Science and Technology, Staatskerant, 8 April 2016. No. 39910 pp. 39–69. Available at http://pmg-assets.s3-website-eu-west-1.amazonaws.com/b_6_-_2016_protection_promotion_development_and_managment_of_indigenous_knowledge_systems.pdf, accessed on 12 June 2019. In South Africa, a Bill is a draft version of a law, and before becoming a law it must be

considered by both houses of Parliament (the National Assembly and the National Council of Provinces). Once it has passed through these houses, it goes to
the President for assent (signing into law); once signed it becomes an Act and
law of the land.

8. Drafted by the Department of Science and Technology (n.d.).
9. The Principle of Prior Rights "recognizes that communities have prior, proprietary
   rights and interests with all knowledge and intellectual property and traditional
   resource rights associated with such resources and their use."
10. For example, the "Informed Consent" requirements assume that if an individual
    consents to sharing knowledge publicly, the knowledge can then be shared.
11. See the United Nations Treaty Collection, Chapter XXVII, Environment, 8.
    Convention on Biological Diversity. Available at https://treaties.un.org/pages/View
    Details.aspx?src=TREATY&mtdsg_no=XXVII-8&chapter=27 accessed on 12 June 2019.
12. When a country signs onto an international treaty, it does not bind the State to
    the provisions within the treaty.
13. *Sui generis* can be defined as of its own kind, and in the intellectual property
    law context describes a regime designed to protect rights that fall outside of the
    traditional patent, trademark, copyright, and trade-secret doctrines (see World
    Intellectual Property Organisation—Glossary available at http://www.wipo.int
    /tk/en/resources/glossary.html#s, accessed on 12 June 2019). Countries are devel-
    oping *sui generis legislation* to specifically address the positive protection of IK.

References

Barume, Albert Kwokwo. 2010. *Land Rights of Indigenous Peoples in Africa: With
    Special Focus on Central, Eastern and Southern Africa*. IWGIA Document
    115. Copenhagen: International Work Group for Indigenous Affairs.

Foster, Laura. 2011. "Situating Feminism, Patent Law, and the Public Domain."
    Colum. J. Gender & L. 20: 262.

Maguire, Patricia. 1996. "Considering More Feminist Participatory Research:
    What's Congruency Got to Do with It?" *Qualitative Inquiry* 2 (1): 106–18.

Makoena, S. 2015. "Mixed Views Heard at Traditional Leadership Bill Public
    Hearings." http://www.parliament.gov.za/live/content.php?Item_ID=8642.

Natural Justice. 2015. "Traditional Knowledge and Customary Sustainable
    Use of Biodiversity: E-learning Series on International Frameworks that
    Support Indigenous Peoples, Local Communities, and Their Territo-
    ries and Areas." http://naturaljustice.org/wp-content/uploads/2015/09
    /Traditional-Knowledge.pdf.

Republic of South Africa. 2004. *Indigenous Knowledge Systems Policy*. South Africa:
    Department of Science and Technology. http://www.dst.gov.za/index
    .php/legal-statutory/policies/1258-indigenous-knowledge-system-policy.

Republic of South Africa. 2016. "Protection, Promotion, Development and
    Management of Indigenous Knowledge Systems Bill." South Africa:
    Ministry of Science and Technology. https://www.gov.za/sites/default
    /files/b6-2016_a.pdf.

Republic of South Africa. n.d. *Draft Indigenous Knowledge Systems Research Ethics Policy.* South Africa: Department of Science and Technology (DST).

Savaresi, Annalisa. 2016. "Doing the Right Thing with Traditional Knowledge in International Law: Lessons for the Climate Regime." BENELEX Working Paper 8, Edinburgh School of Law, 2016/16.

Schonwetter, Tobias, Lesle Jansen, and Laura Foster. 2015. "Submitted Comments on the Draft Protection, Promotion, Development and Management of Indigenous Knowledge Systems Bill 2014, General Notice 243 of 2015." http://naturaljustice.org/wp-content/uploads/2015/03/PPDM-Indigenous -Knowledge.pdf.

Smith, Linda Tuhiwai. 1999. *Decolonizing Methodologies: Research and Indigenous Peoples.* London: Zed Books.

Smith, Laura, Lucinda Bratini, Debbie-Ann Chambers, Russell Vance Jensen, and LeLaina Romero. 2010. "Between Idealism and Reality: Meeting the Challenges of Participatory Action Research." *Action Research* 8 (4): 407-25.

United Nations (UN). 2012. "Indigenous Peoples of Africa Coordinating Committee (IPACC) Statement to the 11th Session of the UN Permanent Forum on Indigenous Issues (UNPFII)." New York: United Nations.

——. 2008. "UN Declaration on the Rights of Indigenous Peoples (UNDRIP)." 07-58681. http://www.un.org/esa/socdev/unpfii/documents/DRIPS_en.pdf.

United Nations Environment Programme (UNEP). 1992. "Convention on Biological Diversity." United Nations. https://www.cbd.int/convention /articles/default.shtml?a=cbd-08.

CHAPTER 11

Negotiating Openness in Science Projects: Case Studies from Argentina

Valeria Arza and Mariano Fressoli

Abstract

Open Science promises to revolutionize the scientific model of knowledge production, and as a result, scientific and funding institutions have increasingly started to adopt its policies. However, most policies are limited to the institutional level, and, in developing countries, there are no models that inform how to build good practices of openness at the laboratory level. This chapter analyzes four cases of Open Science in Argentina, characterizing what is being opened, how, and who participates in these practices. The analysis shows that as scientists open more stages of their research, they enter into a social terrain that challenges their formal scientific norms and customs. We tentatively study this moment through the notion of boundary objects to understand how scientists negotiate meanings, tools, and several forms of communication with actors from outside the laboratory. In the conclusion, we suggest the need to identify and build exemplary cases of Open Science that allow the construction of good practices.

Introduction

Open Science is increasingly gaining attention from scientists and policy makers. Scientific institutions, funding organizations, and policy makers worldwide, such as the OECD (OECD 2015), the World

Bank (Rossel 2016), and the European Union,[1] have demonstrated interest in the practices of Open Science. In Argentina, the Law 26.899 of Open Digital Repositories, in force since 2013, and the trend to foster networked research projects provide an opportunity to adopt the tools of Open Science. Understandably, public policy and institutional recognition of Open Science seem to be focusing on technical areas where there are existing capabilities or it is easier to create them (e.g., David 2004). Therefore, institutional policies have favoured practices such as Open Access and Open Data. However, this initial process of opening up research outputs has not spread through other research stages. This approach is not unique to policy-making institutions. As the few studies about the opening up process suggest (e.g., Whyte and Pryor 2011), normally researchers do not commit to total openness but rather attempt to open up pragmatically. However, it is still not clear what aspects of the research cycle scientists and institutions are choosing to open and how—what negotiations take place?

One problem facing researchers who are inclined to Open Science is that there is no model, necessarily, that can guide them in changing their daily scientific practices. Openness and collaboration with other actors outside of the laboratory (either other researchers or citizens) undoubtedly challenge the adopted norms and customs of traditional scientific work. Also, every stage of the research process faces specific challenges in terms of infrastructure, management, and participation mechanisms, as well as risks of the undue appropriation of results. Some disciplines, such as mathematics, astronomy, and ecology, appear to be advancing more rapidly than others in the above-mentioned process.

This raises questions about the best spaces and strategies to initiate the process of Open Science, about the tools and capacities that need to be developed, and about the challenges faced by practising Open Science in different contexts. One no less important point is that most of the pioneering examples of Open Science, such as the Polymath project, Galaxy Zoo, or Foldit, which have motivated studies about Open Science, originated in universities and networks from developed countries. As the success of Open Science projects depends on factors embedded in specific contexts, these pioneering examples cannot always be directly transferred to other places. This chapter aims to understand how openness is realized in the context of Argentina, a country where the attention to science-related policy has recently grown, but investment still remains low compared with more developed countries.[2] Based on four case studies, which belong to four different networks of knowledge production, we examine what, how, and toward whom the opening process advances: when and why it takes place; what resources are necessary; and what specific capabilities scientists need to develop, and we outline the major lessons and challenges.

In Section 2 of this chapter, we discuss Open Science practices and policies. We argue that there is no clear route to follow to manage the opening-up of scientific initiatives, much less in developing countries. Section 3 presents the conceptual framework and the methodology used to analyze four Open Science projects from Argentina. This analysis is done in Section 4. Section 5 explores whether scientists construct boundary objects in the process of opening up. Boundary objects (Star and Griesemer 1989) are translation devices that connect meanings and practices across different communities. Finally, the conclusions suggest new lines of research and policy action.

Section 2: Practices and Policies of Open Science

New information and communications technologies (ICTs) have provided the opportunity to create open forms of collaboration between scientists in the definition of research problems (for example, in the Polymath project; Nielsen 2012); the participation of citizens in data classification and analysis (for example, Galaxy Zoo, Foldit; Franzoni and Sauermann 2014); or the design of software or scientific instruments for Open Science (for example, the statistical software R or the Geiger counter; Pearce 2012). Scientists are increasingly called upon to share publicly funded research outputs, such as data, publications, and infrastructure. In general, the funding agencies have demonstrated growing interest in promoting the common use of instruments that require significant investments (Sonnenwald 2007). Furthermore, there is a lot of progress in the creation of open repositories for scientific papers, although gradually repositories for data have also been developed (Gagliardi, Cox, and Li 2015).

Diverse international organizations and scientific institutions have begun to carry out recommendations and to put forward policies for the implementation of Open Science practices: for example, to open up datasets (OECD 2015; Stodden 2010), to promote access to systematic data management services (EU 2016), to acknowledge the support of open (software and tools) infrastructure (RIN NESTA 2010; Stodden 2010), and to innovate in scholarly communication practices (EU 2016). Recommendations of scientific institutions and of developmental organizations are oriented toward creating policies at the institutional level, but they offer limited guidance on how to carry out opening projects at the level of project, the laboratory, or the scientific network.

In Argentina, public policy has been almost exclusively focused on Open Access. The country was a pioneer in the region,[3] obtaining specific legislation to guarantee Open Access to publicly funded scientific outputs (through the National Law for the Creation of Digital, Institutional and Open Access Repositories that was approved in 2013 and fully in force since 2016).

However, despite these great advancements in Open Access, there is still little talk on how Open Science can move forward in other aspects of the research cycle (including citizen participation in data recollection, open peer review, public hearings). While enthusiasts from the Open Access movement initially advocated Open Access policies, it is still not clear who will push for Open Science and how scientists are going to engage in the process. As a recent study shows, scientists are not very aware of Open Science practices beyond Open Access, and there is some misunderstanding about the meaning of Open Science, although at the same time there is a great level of interest in making scientific production more collaborative and open (Arza, Fressoli, and Lopez 2017).

The lack of models or guides to follow (RIN NESTA 2010) might also reflect the cautious attitude of scientists toward openness (Whyte and Pryor 2011). At the same time, however, some opening processes can require more negotiation than others. For example, difficulties in using Open Source resources and tools, tensions between the research culture and the processes of opening, and participation of the public (Wylie et al. 2014; Riesch, Potter, and Davies 2013).

There are still a lot of challenges to the practice of Open Science, including individual and institutional obstacles (Sheliga and Friesike 2014). But, while in the European and North American contexts there is an increasing network of institutions (including scientific institutions, as well as NGOs) that offer tools,[4] protocols, and tutorials to help introduce scientists and citizens to the world of Open

Science, this infrastructure is mostly absent in Argentina, where there are neither specific programs nor support for these practices beyond Open Access.

Section 3: Conceptual Framework

In order to analyze the practices of opening the selected initiatives, we began with the characterization of RIN/NESTA (2010) on three relevant dimensions characterizing the openness of the different phases of scientific production, summarized as follows:

1) **What is opened:** This refers to which goods are put into open availability. The Open Access movement traditionally advocated for access to the final result of the scientific production process. More recently, the movements of Open Science have also focused their attention on other types of material and other phases of the research process, such as raw data, curated data, research protocols, laboratory notes, and project proposals.

2) **How is it opened (or which conditions enable the opening):** The grade and scope of openness for intermediate and final outputs of the research process vary according to several restrictions that are made more or less explicitly. These restrictions can be formal, such as the paid subscriptions or licences for the use of material or information (Molloy 2011), or informal, such as the need to obtain certain skills or complementary resources to be able to enjoy the most benefit from shared knowledge.

3) **Who participates or who are the targets of openness:** Scientists are used to sharing the final results of their research with colleagues from the scientific field, but they are less prepared to share their results with a much broader audience. The practices of Open Science have the goal of amplifying the quantity and diversity of the users and producers of scientific knowledge.

Methodology

We performed a case study analysis to understand how the processes of Open Science were carried out in concrete cases. Particularly, we aimed at analyzing the dynamics of the open and collaborative production of knowledge and data in terms of the dimensions of the research cycle that were opened, the timing of openness, the obstacles faced by researchers, and the infrastructure they used. We selected cases from a survey of all researchers working in the national scientific system, taking into account the need to cover the widest possible diversity of situations and opening processes, in terms of disciplines, processes of knowledge creation, techniques of participation, and type of infrastructure used. The selected projects are: New Argentinean Virtual Observatory—NOVA (astronomy); Argentinean Project of Monitoring and Prospecting the Aquatic Environment—PAMPA2 (limnology), e-Bird Argentina (ornithology), and Integrated Land Management Project (geography, chemistry, and environmental science). To gain information on these case studies, qualitative research methods were used, including the review of primary sources (such as scientific papers, reports, newspaper articles, and material available on the web), secondary sources, and semi-structured interviews (twelve in total, three per case), which involved scientists and technicians from the different initiatives.

Section 4: Cases of Open Science in Argentina

In this section, we present our four case studies, describing the origins and motivations of each experience, the development of the infrastructure, opening-up mechanisms, and the outcomes they obtained.

Case Study 1: New Virtual Argentinean Observatory—Nova[5]

NOVA was founded in 2009 with the aim of collecting and centralizing previously processed astronomical data in order to integrate local data to international standards, to allow its reuse, and to promote the development of astronomy. The initiative brings together the most important astronomical research centres in Argentina and counts on the support of the National Science and Technical Research Council (CONICET) and the Ministry of Science, Technology and Productive Innovation (MINCYT). The financial support allows NOVA to hire a software technician and to become part of the International Alliance of Virtual Observatories (IVOA).

NOVA gathers astronomical data in the form of images, spectrums, catalogues, measurement lists, and tables. Originally, much of this data are generated automatically by telescopes and processed later by scientists who integrate it into their analysis. However, after the analysis is done, the data are not usually re-used and sometimes are even forgotten. In addition to that, there is extensive data available in analog pictures or measurements, which are not digitized and which NOVA seeks to recover. For this reason, the aim of the project is to gather the data generated by local scientists and to make it freely available.

Until now, NOVA has mainly gathered a data collection called "Variable View of the Milky Way," which involves about four hundred million space positions. As a virtual observatory, NOVA has not required large investments in terms of infrastructure. The development of the site is based on using existing software, such as open software from the Virtual German Observatory (GADO). The greater investment was to buy a server and some personal computers to save data. Moreover, CONICET pays for the salary of a technician who is in charge of maintaining and updating the database, and of developing software applications and other tools.

Among the tools generated locally is an open software application to automatically upload and validate new pictures. NOVA also developed digital manuals and organized training sessions for astronomers to encourage the use of the NOVA site. From the beginning of 2015 until November that year, the NOVA site had about eighty-five thousand visits, of which one thousand two hundred and thirty-eight were data downloads including those from national researchers as well as researchers from other countries.

Recently, a group of scientists and students from the Laboratory of Research and Formation of Advanced Informatics from the National University of La Plata have started a citizen science initiative using NOVA Open Data. Specifically, they have begun to develop electronic games, which allow the general public to collaborate in the classification of data, such as of galaxies. One of the games allows users to discover new galaxies, which are validated later by scientists. This development is also part of a much larger project called *Cientópolis*, which aims at producing a platform for citizen science, not only for astronomy but also for other endeavours. According to Robert Gamen, director of NOVA: "The experience has been so positive…what began as a game may end up being something about which people will talk for years."

Case Study 2: Argentine Monitoring and Prospecting Project of Aquatic Environments—PAMPA2

PAMPA2 is an interdisciplinary network that seeks to understand the response of the Pampas' lagoon ecosystems to climate variability, changes in land use, and other anthropogenic effects. The central idea is that lagoons can act as "sentinels" that allow for observation of larger changes in the environment. This required a team of interdisciplinary researchers composed mostly of oceanographers, geographers, meteorologists, biologists, zoologists, and engineers to study inland water bodies selected in three provinces over a period of five years.

The network sought to create a long-term monitoring process for thirteen lagoons located along a gradient of decreasing humidity in the provinces of Buenos Aires, Córdoba, and Santa Fe. In five of the thirteen lagoons, buoys equipped with automatic sensors that measure temperature, pressure, wind, rainfall, humidity, oxygen, chlorophyll, and depth were installed. These devices are connected to a processor that stores information and then transmits it in real time to the laboratories of the network.

The data from the buoys are supplemented with other data generated by sampling in the field on a monthly or biannual basis according to the variable selected, both in lagoons that do not have buoys and in those in which buoys are already in place. These data are not open.

Since PAMPA2 is funded by CONICET, a certain level of data access must be offered. In practice, this means free availability to data produced by some of the buoys in real time (which can be accessed by anyone) and the possibility of access to bigger data sets (which generally are requested by scientists). The project does not yet have any standardized protocol on data access, although this is a current issue on the agenda of the research team.

The IADO develops and produces most of the instruments, including the automated environmental monitoring buoy in hydrology and most of the integrated sensors. In 2011, the buoy won second place in a national Innovation Award. Currently, researchers at IADO are working on a new version of the buoy that will use Open Source software. They seek to give the project an international scope and to add the collaboration of other stakeholders. The creation of PAMPA2 has enabled an increasing interaction with similar research projects around the world. PAMPA2 integrates GLEON Network (Global Lake

Ecological Observatory Network), an organization of global institutions that monitors lakes steadily through instrumented buoys. This network aims at standardizing the format of data obtained by buoys from eighty different locations, but the members have not yet reached a consensus on what database system they will use.

One of the team leaders of PAMPA2 whose group is currently involved is the SAFER project (Sensing the Americas' Freshwater Ecosystem Risk from Climate Change), an initiative that integrates scientists from various specialties from Argentina, USA, Canada, Chile, Uruguay, and Colombia that uses community-based strategies to produce knowledge. The diffusion of results to a wider audience is contemplated among the goals outlined by SAFER. For instance, this implies plans to spread the results of the project among the populations in the vicinity of the lagoons. However, diffusion activities have not been carried out so far because of the lack of technical and financial resources. Another obstacle is that the website that shows the data generated by the network is under construction and is not designed to receive queries from the public. Yet, researchers receive regular inquiries from people who consult the data available, for purposes such as recreation and/or production. According to Gerardo Perillo from PAMPA2:

> People who know that it exists and that is getting access to data that has not existed before... . To those the project has helped... they could find the data useful. The only weather station from Monte Hermoso, or Pehuen-có is our station, so they enter our station to know what data are available... . But we also have to be cautious: it is something that we do and we release freely available but these are research stations, they are not official stations of weather forecast established by an authorized body.

In this sense, as the process of opening of PAMPA2 advances, new challenges have arisen in diffusion of data, which in turn require improved infrastructure and precautions around the use of this data.

Case Study 3: Integrated Land Management Project

The Integrated Land Management (ILM) project is an interdisciplinary project that sought to study the vulnerabilities of two areas affected by severe floods in 2013 in collaboration with neighbours and institutions. These areas are the basin of the Maldonado Stream and that near a large oil refinery in Ensenada and Berisso, in the province of Buenos Aires. The aim of the project was to assess the environmental and social consequences of the floods and to propose solutions. The project was led by a group of social scientists and environmental chemists.

The project had two phases: diagnosis and implementation of solutions. At the time of our interviews, the team was well into the first phase, which aimed at doing a systematic assessment of environmental and social problems that the community recognized and required to be solved. To that end, the team articulated various techniques of natural sciences with methods of intervention from the social sciences. The expectation was that combined results from these different methods would allow the design of some solutions to existing problems, which were going to be implemented in the second stage with the partic- ipation of neighbours, institutions, scientists, and companies. This research went side-by-side with the development of technological solutions by the team of environmental chemists.

Citizen participation took part in several stages: during the de- sign of the survey form; in the collection of rainwater to measure the pH level in order to detect the acidity or alkalinity of water; in the identification of patterns of territorial appropriation at the micro level; and in the discussion of concrete actions of intervention, among others. The analysis of all collected data was then processed and interpreted by researchers (without the participation of the neighbours).

The research outcomes have been incorporated into the repos- itory at La Plata Environmental Observatory (OMLP). However, re- searchers claim that the dissemination has to be done with caution to avoid alarming or causing a negative impact on the population's beliefs and on the local authorities.

Similarly, researchers must be cautious regarding the manage- ment of neighbours' expectations since they cannot guarantee that the proposed solutions will actually take place. On their side, neighbours are also cautious about their degree of commitment to the project since this was not the first project that required their collaboration, without always delivering the expected solutions.

These precautions are illustrative of the difficulties and continu- ing renegotiation that a community engaged in Open Science projects must endure in order to open the research and results to a wider public. On top of this, there are further issues to be negotiated that have to do with the political context, as this is a project that is well embedded in the local authority policy agenda. For example, as there were local elections and the ruling party in the local government changed in the middle of the project's timeline, researchers needed to negotiate with the new authorities regarding what each party was expected to deliver, who in turn had to obtain approval from the neighbours.

Case Study 4: E-Bird Argentina

E-Bird is a citizen science project that receives bird sightings from anyone in any part of the world through a website and mobile phone applications. The project builds on the tradition of observation, photography, and bird conservation dating back, at least, to the late nineteenth century. It is an online platform developed in the United States in 2002 by the Ornithology Laboratory at Cornell University, which then expanded its scope, incorporating local partners in different countries. In Argentina, e-Bird was launched by the non-governmental organization Aves Argentinas in 2013. For the project, Aves Argentinas depended on the support of a network of eighty bird watching clubs. The website is maintained with the supervision of the technical staff at the institution, which has also the task of promoting and training users.

To adapt the portal for local use and launch it, Aves Argentina requested public funding, used partly in the implementation of training courses in birdwatching. E-Bird is built on the simple concept that whenever an observer grabs a pair of binoculars, he/she has the opportunity to gather useful information about the occurrence of species, migration time, and the relative abundance in a variety of locations and times. E-Bird makes use of the internet as a tool to collect, archive, and distribute information efficiently to a much wider audience.

Birdwatchers that use e-Bird to report their observations should follow a standardized protocol to load their data to ensure consistency and quality of records. Data uploaded by the users is checked in turn by a series of semi-automated mechanisms. In the case of unusual uploaded data, these are reviewed by a designated expert who controls its veracity. In Argentina, in addition to the four people who work for Aves Argentinas, twenty amateur experts collaborate in data verification.

Every e-Bird local portal is integrated within the infrastructure of applications and the database located in the United States. Despite this centralization, e-Bird is an open platform. This allows, for example, any user to have access to simple data from the website. In the case of large volumes of data, it can be requested for free from e-Bird in the US, and the data are returned by email. In addition, Aves Argentinas and its funders made an agreement to join the National Biological Information System (SNDB) that involves a commitment to incorporate data from e-Bird to SNDB.[6]

Data gathered by e-Bird that provides information on the spatial distribution of species and allows the possibility of tracking population trends, also help in identifying areas or important sites for the conservation of birds. Thus, e-Bird might contribute to the design of better management plans for the recovery of threatened species or for those in danger of extinction. At the same time, these data can be used for scientific purposes to study the distribution patterns and movement of birds throughout Argentina, including migration routes, wintering and breeding areas, etc. At this time, it allows amateur observers to know more about birds in the region they inhabit and assists in tracking their personal observations.

In little more than two years of operation, the e-Bird Argentina project achieved the detection of approximately nine hundred and sixty-seven thousand different species, which is approximately ninety-five percent of the species that exist in Argentina. It is likely that this collection would not have been possible without the participation of hundreds of enthusiastic citizens who contributed their data.[7]

Characteristics and Scope of Openness

Following the concepts presented in the introduction, in this section we look to understand the characteristics of the process of openness, how it has evolved, how obstacles are overcome, and which stages are opened and why.

What Is Being Opened: Data, Infrastructure, and Citizen Participation

The four cases have the common goal of opening data for re-use by scientific networks and by citizens—although they have had different results in doing so. In the case of NOVA and PAMPA2, the release of data is mainly based on the international practices of their respective disciplines. Part of the incentive of opening up these cases is the ability to share data and research on a reciprocal basis with researchers and international institutions. In the case of e-Bird, the incentive for opening is different because data producers are not scientists, but citizens—thus, opening works as an incentive to share data in a community of peers, even if researchers in various disciplines also use the data. In these three cases (NOVA, PAMPA2, and e-Bird), institutional support in the data opening process was provided mainly by their public funders, without the need for an imposition of a plan as to how the data should be released. In the case of ILM, the situation is reversed since there is no obligation to open the data. Although as part of the Environmental Observatory of La Plata, the data would eventually be public, but, at the time this research concluded, data was not yet made open.

A second point in the opening process is infrastructure, in particular, open software. Both NOVA and e-Bird Argentina took advantage of existing open software and made local adaptations using minimal resources. In the case of PAMPA2, researchers took advantage of the expired patents for the assembly of the first monitoring buoys. Later, as it was time to advance a design for new buoys, the use of open software began to be considered as a way of improving collaboration and for resolving problems.

The third focus of openness is the citizen participation in the collection of data. In e-Bird, citizen science constitutes the basis of the project. In contrast, in ILM, the citizens helped to collect some of the data regarding water quality and also to refine the questionnaires, as well as suggesting the best locations for the research. In the other cases, citizen science tools were used only once the project had begun. In NOVA, the opening to citizen participation took place in the context of an informatics workgroup, created within the university that led NOVA, called *Cientópolis*, whose objective was to create a platform for the development of citizen science projects. Similar to Galaxy Zoo (Franzoni and Sauermann 2014), *Cientópolis* has built electronic games, such as the Galaxy Conqueror, that allow users to classify galaxies. PAMPA2 does not experiment with tools for citizen science data collection, but its associated project, SAFER, does. This project has an educational component and works with students from a middle school who collect data to help the research team.

How It Is Opened: Participation and Barriers to Access

The conditions under which the opening process takes place vary according to the objectives and the requirements of data production for each initiative. In the case of NOVA and PAMPA2, data are mostly produced by scientists and for scientists. Therefore, the opening protocol establishes a period of embargo on new data that can last until the publication is complete. However, once this embargo period is over, data are made freely available for use and analysis by other researchers. Nonetheless, in the case of PAMPA2, some of the data obtained during the day can be observed for free on the project website. E-Bird also offers Open Data to the general public on a large scale. However, similar to PAMPA2, the use of large datasets are granted by the website administrator only upon request.

Some of the available data are simple and do not require prior knowledge to make the most of them (PAMPA2 and e-Bird). In the case of NOVA, access to data is free, but requires expert knowledge of astronomy and specific software tools used by the project. The development of the game Galaxy Conqueror seems to aim at alleviating this barrier, at least partially, making data available to allow greater interaction with the public. In the case of ILM again, the conditions for access to the data are limited due to the complex political situation of the floods in the region and the fear that this information could trigger false expectations among the public. Indeed, this last case suggests that the negotiations of openness in the case of politically sensitive information are more complex and mediated differently than other scientific projects.

For Whom It Is Opened: Uses and Benefits

The four cases have implemented some form of Open Access that eventually would allow data to be re-used by other scientists. However, there is little evidence that this is happening at the local level, in contrast with international cases. For instance, in the case of e-Bird, data available from Cornell's servers have been used by researchers in various disciplines, including landscape, ecology, macro-ecology, computer science, statistics, and human computation. Data from NOVA have also been shared at the international level, but so far there is no track of papers published using the Argentine data. In PAMPA2, although some difficulties remain in collaborating across different disciplines, the group has published jointly, including a special journal issue.

Besides scientific collaboration, the four cases show different degrees of openness to public participation as users and/or producers of data. The clearest case is, of course, e-Bird, in which the public plays an important role in data collection and is also a key user of data. Similarly, ILM has involved the public in certain aspects of the research cycle like data collection and questionnaire design. In turn, NOVA (*Cientópolis*) and PAMPA2 (SAFER) are also making efforts to involve the public in the use and production of data. Because this level of openness is under construction in both cases, it is difficult to say how participation will be promoted; it might require the development of new infrastructure (i.e., processes, data validation, and use of social networks more intensively).

Section 5: Negotiating Openness Through the Construction of Boundary Objects

The cases allow us to understand how scientists in Argentina take advantage of the scarce available support from policies and programs in order to explore new forms of openness in other stages of the research cycle. Thus, the opening process is not limited to Open Access and collaboration among scientists from a project and/or discipline, but it is slowly opened to other forms of collaboration with scientists and the public in general. This tendency hints that there might be great potential to extend the practices of Open Science in the country. At the same time, we noticed that opening attempts are gradual and differentiated by the stages of the research process. In this sense, these cases also offer some insights into the limitations and challenges that local scientists suffer when trying to open other stages of the research cycle, due to the lack of tools and the capabilities available for such tasks.

In the cases analyzed, the opening process does not follow an established plan; some of the practices of openness are created in the making. More importantly, as scientists open their data and tools to collaboration with other actors in society, they begin to enter a field that is not always familiar and that can challenge the rules and customs of scientific practice. In this sense, the negotiation phase of the opening process is similar to the construction of boundary objects (Star and Griesemer 1989). This notion was originally developed by

Star and Griesemer (1989) to understand how scientists, conservationists, and amateurs translate different ways of producing information at the Berkeley Zoological Museum of Science. Extending the original use of this concept a little, in the following sections we explore briefly how scientists build boundary objects to negotiate opening on three levels: tools and infrastructure, data to other experts, and communication and dissemination.

Tools and Infrastructure

Opening access to data and the process of collaboration often requires building new infrastructure and technical tools such as software, databases, web pages, and sensors. In practice, this means contacting experts from other areas and communities who respond to quite different aims and rules (such as software programmers, makers, etc.). In two of the analyzed cases, it was possible to see how building these elements was made easier by the availability of open software tools (e.g., NOVA, e-Bird). In the case of PAMPA2, they have recently started to build a new instrument using open software. However, this presents some challenges since the scientists do not always have the capabilities to use and develop this kind of tool. They sometimes have to learn the basics about Open Source software, create new data protocols for Open Access, and begin to understand what data can be made public and what cannot. Beyond the need to develop these capabilities, scientists do not always have the required resources and technical support to develop basic tools such as a web page. Therefore, some of the advances in the process of opening up science are often done *ad hoc* and based on the goodwill of scientists.

Collection and Opening of Data

Similar to the description by Star and Griessemer (1989), standardization and simplification of data, such as the construction of simple forms of visualization, are key tools that allow the use of data by other actors. The same applies to the processes of data collection by citizens, where the development of simple protocols is essential to facilitate public participation. In the case of SAFER (PAMPA2 sister project) and ILM, inviting public participation required the construction of a minimum instrument, and in the case of e-Bird and *Cientópolis* (NOVA sister project), recreational development tools like games and quizzes. Translating the data collection and use into accessible formats can be seen as a challenge. This implies negotiating different meanings and uses of the data with potential users. However, these are also areas of expertise that are rare in scientific labs and scientific institutions but need to be considered in Open Science plans at the institutional level.

Communication and Diffusion

Project visibility is needed to motivate participation and collaboration of diverse actors (e.g., Benkler, Shaw, and Hill 2015). Inviting them to collect data or to collaborate in the design of instruments often requires participatory techniques and communication strategies such as the use of social networks (Lasky 2016). Again, to do this, scientists need to build skills or learn from experts who do not necessarily belong to their scientific field and who are not funded by scientific funding schemes. NOVA has done so at the expense of personal efforts of one of its leaders, who is active on social networks. In turn, e-Bird relies on the international recognition of the initiative and its experiences in organizing competitions, day fairs, etc. PAMPA2 and ILM claimed not to have the resources to do so, although at least the former openly stated they believe it is an important activity.

The central point is that the construction of boundary objects introduces scientists to new fields: (1) relational fields that allow interaction with scientists in other disciplines and with the general public; (2) a technological field that facilitates the development and use of new open technologies; and (3) a management field that allows the coordination of several activities and actors participating in Open Science projects. In these new fields, scientists constantly need to negotiate their knowledge, capabilities, and actions. This negotiation varies across the different fields and also within activities in each of them. It is likely then that the learning processes required to build the necessary boundary objects to enter new fields include not only the accumulated skill sets of scientists, but also their learning capacities to conquer the new tools of open infrastructure, public communication, and management of social networks.

Conclusions

Open Science policies can benefit from the study of exemplary cases. We believe it is important to systematize benefits, challenges, and obstacles experienced by different Open Science initiatives in Argentina. This can help with the creation of an action plan for initiatives that are keen to join the Open Science caravan. As we have seen, the opening process is usually progressive and diverse. It is therefore essential to have a variety of cases that develop a set of good practices. The study of the construction of boundary objects can help in understanding how scientists learn to negotiate their interests and practices during the opening process. In particular, it is important to note that the further scientists engage in the opening process, the more capabilities and tools they will need. Scientific institutions and policy schemes are currently providing neither one. Policy makers might need to consider better policy design to promote Open Science.

Notes

1. See the Open Science Policy Platform set up by the European Union since 2016 at https://ec.europa.eu/research/openscience/index.cfm?pg=open-science-policy-platform.
2. The policy area on Science, Technology, and Innovation reached the rank of a Ministry in 2007. It was previously managed by a secretariat dependent on the Ministry for Education. The higher rank in the Government Organization Chart was highly symbolic, and it correlated with a switch in the policy and media discourse promoting science, technology, and innovation as necessary tools for development. The national spending on R&D has also increased continuously both per capita and as a percentage of GDP, at least until 2012, when economic recession became evident. Both indicators climbed from USD 40.8 and 0.46% in 2007 to USD 90.26 and 0.64% in 2012. In 2014 (latest data available), they were USD 80 and 0.59%. These figures are among the largest in the region, only surpassed by Brazil (USD 147.1 and 1.2% in 2013), but they are quite low when compared with those from United States (1443.9 USD and 2.73% in 2013) or even Spain (342.6 USD and 1.2% in 2014).
3. Peru was the only country in the region with similar legislation, also approved in 2013. The national repository there is called National Open Access Repository of Science, Technology and Innovation (*Depósito Digital Nacional de Ciencia, Tecnología e Innovación de AccesoAbierto*). http://alicia.concytec.gob.pe/vufind/.
4. For a brief guide of available tools for Open Science, see https://www.cientopolis.org/herramientas-de-ciencia-abierta/.
5. The case study of NOVA is based on the work by Rodriguez, F. (2015). Nuevo Observatorio Virtual Argentino—NOVA, in Arza, V., and M. Fressoli (ed.), *Proyecto: Ciencia abierta en Argentina: experiencias actuales y propuestas para impulsar procesos de apertura*. Retrieved from: http://www.ciecti.org.ar/wp-content/uploads/2016/09/CIECTI-Proyecto-CENIT.pdf.

6. However, there are interoperability issues that have hampered the process of migration of Argentinean data from the e-Bird server in the US to SNDB's servers in Argentina. Aves Argentinas is searching for a technical and/or managerial solution to this problem.
7. Globally, the volume of data collected by e-Bird increased exponentially in a period of ten years, 30–40% annually between 2003 and 2013 (Sullivan et al. 2014). By mid-2013, one hundred and forty million observations were collected from one hundred and fifty thousand separate observers, who spent 10.5 million hours collecting data (Sullivan et al. 2014).

References

Arza, Valeria, and Mariano Fressoli. 2016. *Ciencia abierta en Argentina: experiencias actuales y propuestas para impulsar procesos de apertura* [Open science in Argentina: Current experiences and proposals to promote processes opening]. Buenos Aires: Centro Interdisciplinario de Estudios en Ciencia, Tecnología e Innovación (CIECTI). http://www.ciecti.org.ar /wp-content/uploads/2016/09/CIECTI-Proyecto-CENIT.pdf.

Arza, Valeria, Mariano Fressoli, and Emmanuel Lopez. 2017. "Ciencia abierta en Argentina: un mapa de experiencias actuals." *Ciencia, Docencia y Tecnología: Revista científica de la Universidad Nacional de Entre Rios* 28 (55). http://www.pcient.uner.edu.ar/index.php/cdyt/article /view/242.

Benkler, Yochai, Aaron Shaw, and Benjamin M. Hill. 2015. "Peer Production: A Form of Collective Intelligence." In *Handbook of Collective Intelligence*, edited by Thomas Malone and Michael Bernstein. Cambridge, MA: MIT Press.

David, Paul A. 2004. "Towards a Cyberinfrastructure for Enhanced Scientific Collaboration: Providing its 'Soft' Foundations May Be the Hardest Part." Working Paper, Social Science Research Network Working Paper Series. http://papers.ssrn.com/sol3/Delivery.cfm/SSRN_ID1325264 _code1148612.pdf?abstractid=1325264&mirid=1.

European Union (2016) Open Science Policy Platform.

Franzoni, Chiara, and Henry Sauermann. 2014. "Crowd Science: The Organization of Scientific Research in Open Collaborative Projects." *Research Policy* 43 (1): 1–20. http://doi.org/10.1016/j.respol.2013.07.005.

Gagliardi, Dimitri, Deborah Cox, and Yanchao Li. 2015. "Institutional Inertia and Barriers to the Adoption of Open Science." In *The Transformation of University Institutional and Organizational Boundaries*, 107–33. Higher Education Research in the 21st Century Series. Rotterdam: Sense Publishers. https://www.escholar.manchester.ac.uk/ uk-ac-man-scw:283336.

Lasky, Joshua. 2016. "NASA's Juno Mission Is a Case Study in Social Media Excellence." *Medium* (Media/Technology). https://medium.com/digital

-trends-index/nasas-juno-mission-is-a-case-study-in-social-media
-excellence-1bfe2f3ac6b4#.c5o2ylnqo.

Molloy, Jennifer C. 2011. "The Open Knowledge Foundation: Open Data Means Better Science." *PLoSBiol* 9 (12): e1001195. http://doi.org/10.1371 /journal.pbio.1001195.

Nielsen, Michael. 2012. *Reinventing Discovery: The New Era of Networked Science.* Princeton, NJ: Princeton University Press.

OECD. 2015. "Making Open Science a Reality." http://dx.doi.org/10.1787 /5jrs2f963zs1-en.

Pearce, Joshua M. 2012. "Building Research Equipment with Free, Open-Source Hardware." *Science* 337 (6100): 1303–1304. http://doi.org/10.1126 /science.1228183.

Riesch, Hauke, Clive Potter, and Linda Davies. 2013. "Combining Citizen Science and Public Engagement: The Open Air Laboratories Programme." *Journal of Science Communication* 12 (3).

RIN NESTA. 2010. "Open to All?" http://www.rin.ac.uk/our-work/data -management-and-curation/open-science-case-studies.

Rossel, Carlos. 2016. "The World Bank Open Access Policy, 3–5."

Scheliga, Kaja, and Sascha Friesike. 2014. "Putting Open Science into Practice: A Social Dilemma?" *First Monday* 19 (9): 1–14. http://firstmonday.org /ojs/index.php/fm/article/view/5381/4110.

Sonnenwald, Diane H. 2007. "Introduction Scientific Collaboration: A Synthesis of Challenges and Strategies." *Annual Review of Information Science and Technology* 41 (January): 643–81.

Star, Susan L., and James R. Griesemer. 1989. "Institutional Ecology, 'Translations' and Boundary Objects: Amateurs and Professionals in Berkeley's Museum of Vertebrate Zoology, 1907–1939." *Social Studies of Science* 19 (3): 387–420.

Sullivan, B.L., Aycrigg, J.L., Barry, J.H., Bonney, R.E., Bruns, N., Cooper, C.B., Damoulas, T., Dhondt, A.A., Dietterich, T., Farnsworth, A., Fink, D., Fitzpatrick, J.W., Fredericks, T., Jeff Gerbracht, J., Gomes, C., Hochachka, W.M., Iliff, M.J., Lagoze, C., La Sorte, F.A., Merrifield, M., Morris, W., Phillips, T.B., Reynolds, M., Rodewald, A.D., Rosenberg, K.V., Trautmann, N.M., Wiggins, A., Winkler, D.W., Wong, W.-K., Wood, C.L., Yu, J., Kelling, S., 2014. The eBird enterprise: an integrated approach to development and application of citizen science. Biol. Conserv. 169: 31–40. https://www.sciencedirect.com/science/article/abs/pii /S0006320713003820.

Stodden, Victoria. 2010. "Open Science: Policy Implications for the Evolving Phenomenon of User-led Scientific Innovation." *Journal of Science Communication* 9 (1): 1–8.

Whyte, Angus, and Graham Pryor. 2011. "Open Science in Practice: Researcher Perspectives and Participation." *International Journal of Digital Curation* 6 (1): 199–213. http://doi.org/10.2218/ijdc.v6i1.182.

Wylie, Sara A., Kirk Jalbert, Shannon Dosemagen, and Matt Ratto. 2014. "Institutions for Civic Technoscience: How Critical Making is Transforming Environmental Research." *The Information Society* 30 (2): 116–26. http://doi.org/10.1080/01972243.2014.875783.

SECTION 4

OPEN SCIENCE FOR SOCIAL TRANSFORMATION

INTRODUCTION

Halla Thorsteinsdóttir

To explore the potential contribution of open and collaborative science for social transformation, this section presents case studies of initiatives pursuing open and citizen-based science in Africa, Asia, and Latin America. The goal of this work is to apply the Open Science approach to a development context and to question how or if Open Science might contribute to positive societal impacts and, on a larger scale, transform the way that knowledge is valued and legitimized in an unequal global context. The three case studies highlighted in the chapters of this section employ different methods and tools to explore this core theme.

Chapter 12 by Rosset et al. focuses on the results of a citizen science project in post-Soviet Kyrgyzstan, which centred on the testing of water quality in rural villages. The project worked specifically with schoolchildren and their teachers in an under-resourced, mountainous areas of the country. The case study on open and collaborative science by Albagli et al. (Chapter 13) focuses on the Ubatuba municipality, a coastal community in the state of São Paulo in Brazil. The municipality has a mixed population, including both powerful and marginalized actors, all competing to make their voices heard in regard to how the region's vulnerable ecosystems should be used and/or protected. Finally Chapter 14, presents the third case study, which focuses on the higher education sector in Francophone Africa (Benin, Burkina Faso, Cameroon, Chad, Democratic Republic of the Congo, Gabon,

Ivory Coast, Madagascar, Mali, Niger, and Senegal) and Haiti. It identifies obstacles to Open Science in French-speaking Africa and Haiti and explores approaches for universities in the region to serve local sustainable development to a larger extent. While the Brazilian and Kyrgyzstani case studies follow a flow similar to the other sections of the book, the final chapter—set in the context of Francophone Africa and Haiti—provides valuable lessons with regard to the "cognitive injustices" that limit the extent to which Open Science can be applied as a useful framework within the region's universities.

The three case studies use a variety of methods, all of which ground their work in a paradigm of community-action research. The diversity of methodologies includes interviews, questionnaires, workshops, social media discussions (particularly through Facebook and WhatsApp), engagement through science-related local events, community forums, and others. Whereas the Kyrgyzstan case study involves hands-on Open Science activities, the Ubatuba and Francophone Africa/Haiti projects have primarily sought to initiate critical reflection and discussion on the concept of Open Science by various actors in their respective regions. All three of the case studies use an educational component on one hand, but also encourage community members to actively take part in agenda-setting for the research and data collection.

Open Science is a relatively new concept for the communities involved in these case studies, although Open Access has become relatively well-understood in the context of Africa's Francophone universities. In the Kyrgyzstan project, the concept was initially met with significant resistance due to the post-Soviet political culture, which remains suspicious of citizen engagement—whereas in Brazil the concept of Open Science was quickly adjusted to "Community Science," to better suit the needs of the community members with whom the team engaged.

The various Open Science initiatives are facilitated as empowerment opportunities for local populations, and some observations point to this being particularly beneficial to populations of lower socioeconomic status. This implies that Open Science provides a unique opportunity to facilitate access to the creation and dissemination of local knowledge, often inaccessible to marginalized populations. If harnessed effectively, it may contribute to political empowerment and mindset transformation regarding how and by whom knowledge should be created.

All three chapters in this section reported findings which suggest that those who engage in reflections around how knowledge is created, shared, and legitimized might begin to shift their ideas on the importance of science, and the creation of locally relevant knowledge, for sustainable development. The case studies emphasize that in order for Open Science to be a seed for change, there is a need for learning and genuine participation in science activities to take place in those contexts where marginalized groups are often excluded from knowledge-production processes. The participation of community members in Open Science initiatives should thus not be limited to data collection efforts, but must rather be extended to planning the research questions, agendas, and methodologies, as well as analyzing and communicating collected data. The authors also make the case that the participation of communities leads to more locally relevant data.

As other chapters have indicated, science and knowledge-production processes have traditionally tended to be exclusionary and conducted in a hierarchical fashion. The case studies in this section demonstrate that those who have been traditionally excluded from such processes can also be enthusiastic about science and report positively about their experiences taking part in such initiatives. However, mistrust and unequal power relationships between scientists and local communities are hindrances to the type of collaboration that could lead to larger-scale transformation.

In that regard, in order to take advantage of the potential of Open Science, there are a number of challenges that need to be considered. The case study of Francophone Africa and Haiti, for example, highlights nine "cognitive injustices" that need to be addressed in order to foster truly open and collaborative science. For instance, the authors suggest that digital literacy and access to the Internet are rare throughout the region, even within some universities, and the most marginalized populations experience the most significant "digital divide." This indicates the need to look at Open Science from a systemic perspective, and to map the key actors and conditions that need to be involved for science activities to lead to innovation and transformation. Particular attention needs to be paid to possible misalignments that can limit the necessary knowledge flow between actors, which could in turn hinder the potential for transformation (Lundvall et al. 2009).

Efforts to Strengthen Impacts of Open Science

While Open Science initiatives on a smaller scale, such as those reported in this section of the book, can demonstrate positive societal impacts, it can be challenging to scale these up to contribute to larger social transformation. Such small-scale initiatives tend to be driven by a few individuals who believe open and collaborative science (OCS) has beneficial societal impacts and who are compelled to promote it. In order for Open Science to contribute to more large-scale societal transformation, work promoting it has to be carried out at multiple levels—including at varying levels of government (municipal, regional, and national), as well as by community organizations and regional institutions.

Various tools that support Open Science for development can be used, including educational modules aimed at communities, students, teachers, scientists, and others. The modules should be based on research on Open Science approaches and be cognizant of the culture of the communities involved. All three case studies in this section demonstrate that mistrust and unequal power dynamics (including within universities and between scientists and communities) are challenges to the expansion of Open Science. There is hence a need for research to look deeper into these issues and work with particular communities in developing guidelines that promote more equal power relationships in Open Science initiatives—such as the community-researcher contracts articulated by Traynor, Foster, and Schonwetter in a South African context (see Chapter 10.)

Communication between traditional knowledge makers and communities is another factor that is often problematic, as noted by all three case studies, in supporting a climate for a fair and equitable Open Science. It is therefore necessary for all actors to learn to respect and listen to each other and to be able to articulate their ways of working in lucid and coherent ways. It is also important to adjust models of communication to match the technologies (or lack thereof) to which particular communities have easy access.

Whichever strategy is relied upon in promoting OCS, it is important to look at the science from a systemic perspective and understand what actors, factors, and conditions are shaping its development. This section provides the background for understanding the potential of Open Science to change the way that development is understood and achieved by transforming how ordinary people understand and participate in processes of knowledge creation in order to facilitate locally appropriate and sustainable change.

References

Lundvall, Bengt Å, K. J. Joseph, Cristina Chaminade, and Jan Vang. 2009. *Handbook of Innovation Systems and Developing Countries*. Cheltenham, UK: Edward Elgar.

CHAPTER 12

Experimenting with Openness as a Seed for Social Transformation: Linking Environmental Education and Citizen Science in Remote Mountain Villages of Kyrgyzstan

Aline Rosset, Aliya Ibraimova, Aikena Orolbaeva,
Altyn Kapalova, and Bilimbek Azhibekov

"Citizens and kids merely doing the grunt work of coming up with data is not the point of citizen science. The point is to engage them in inquiry-based learning and stewardship of the environment."
—Karen Matsumoto

Abstract

In post-Soviet Kyrgyzstan, "science," as understood by most citizens, consists of highly technical and expensive activities performed by scientific "experts." The Kyrgyz Mountains Environmental Education and Citizen Science (KMEECS) project has sought to challenge these widely held assumptions by engaging rural schoolchildren and their teachers in biological, chemical, and physical analyses of water quality, as well as water flow measurement and mapping of locally relevant water resources. Using a participatory action research approach, this project looks at the transformational potential of citizen science initiatives for environmental monitoring and education. It also provides insight on the motivational factors for citizen science at the local level and the complexities of collaboration and support between community and governmental institutions in a post-Soviet state.

Introduction: Open Science and Environmental Education in the Mountains

Open and Collaborative Science (OCS) in development, including citizen participation in scientific research, encompasses approaches that are not widely used in post-Soviet Central Asia, although these approaches arguably offer opportunities to impact education, citizen awareness, policy, and local governance for a more informed management of natural resources. The Kyrgyz Mountains Environmental Education and Citizen Science (KMEECS) project operates in a climate of emerging discourse around openness, sharing, and collaboration in Central Asia. This chapter aims to contribute to the dialogue on the applicability of Open Science (and, in particular, citizen science) for social transformation in Kyrgyzstan, balancing educational, scientific, societal, and policy goals.

Citizen science, as one manifestation of Open Science, potentially enables people to engage with science on real-world environmental issues, in collaboration with scientists working in local contexts (Cohn 2008; Shirk et al. 2012; Bonney et al. 2014). As highlighted by Dickinson et al. (2012), ecological data collected through citizen science can be viewed as a public good that is generated through increasingly collaborative tools and resources. Public participation in science is also regarded as a critical component of "Earth stewardship" (Chapin et al. 2011). Human activities affect Earth's life support systems so profoundly as to threaten many of the ecological services that are essential to society. To address this challenge, a new science agenda is needed that integrates humans within nature to help chart a more sustainable trajectory for the relationship between society and the environment. Similarly, in Kyrgyzstan, an increased awareness of the environment by younger generations is important, since they will be primarily responsible for managing the natural resources under changing environmental conditions.

One quarter of the Earth's terrestrial surface is covered by mountains, which provide goods and services—such as the provision of clean, fresh water—to more than half of humanity. Remote mountain regions are often poorly connected to infrastructure and services. This is also reflected in difficult access to information and knowledge relevant for sustaining local livelihoods under changing socio-ecological conditions. At the same time, environmental monitoring and scientific research are challenging and costly to conduct in remote areas, whereas data and information on the specific environmental, social, and economic assets, as well as the challenges of such regions are often unavailable. Such disparities between urban and rural regions in Kyrgyzstan[1] are also reflected in the education system. Poorly equipped schools, a lack of teacher training, and low salaries lead to reduced motivation and fewer possibilities for conducting hands-on and interactive teaching. Science subjects are taught only in theory due to a lack of resources, contributing to a low interest in science among students.

Citizen science, involving non-scientists in the planning and conduct of research, has often been named as a suitable tool for introducing applied field teaching into theoretical curricula, enhancing student knowledge and involvement with their environment, and, at the same time, contributing to the generation of scientific data (Gommermann and Monroe 2012; Buytaert et al. 2014). Although participatory or citizen science is not a new phenomenon, the past decade has seen a rapid increase in the number of citizen science projects, particularly in North America and Europe, spanning diverse areas of interest and ranging from local to global (Silvertown 2009; UK-EOF 2011; Dickinson et al. 2012; Bonney et al. 2009; Nov, Arazy, and Anderson 2011; Mackechnie et al. 2011; Roy et. al. 2012). However, to date very few citizen science projects are being implemented in developing countries. Similarly, the combination of citizen science and education is not new in the scientific literature, but, to date, it has not been researched extensively in countries of the Global South and even less in the high-altitude and remote rural areas of Central Asia.

The term "citizen science" remains fuzzy and contested, covering a variety of participatory scientific activities balancing educational, scientific, societal, and policy goals (OECD 2015). Depending on the project, the level of involvement of citizens varies, ranging from computer-based crowdsourcing to citizen-designed research. It has been argued that citizen science is a means for reaching several different objectives, as a win-win approach, where a project simultaneously delivers public engagement and scientific research. Therefore, citizen science is seen to have the potential to foster social transformation through the active communication of scientific information needed to initiate a public dialogue and empower people to take ownership of their local environment (Riesch, Potter, and Davies 2013). Taking into account the breadth of definitions of citizen science, the different degrees of collaboration, the variety of participants, as well as the high expectations related to the concept, there are concerns about the ability of citizen science to effectively overcome the many challenges that are apparent in many traditional knowledge-production processes. In particular, when it comes to inclusiveness and the barriers to participation, analyses of participants in some citizen science projects have shown a replication of disparities (Haklay 2012) with regard to educational level, wealth, global geographical distribution, as well as remoteness and accessibility.

Social transformation rarely starts as a large-scale movement. It often starts from "seeds." Although the KMEECS project operates locally, it involves a variety of partners in rethinking the role of science, education, the environment, and civic action. By working with remote mountain communities, the project also raises the topics of remoteness and disparities across the rural-urban development gap. And it plants a seed for an openness movement in Kyrgyzstan by initiating a dialogue between rural activists, teachers, students, policy makers, and scientists.

Based on the KMEECS project as a case study, this chapter discusses citizen science as implemented in a local-level, grassroots project. The next section introduces the case study and the local context, while the subsequent section presents findings in relation to five dimensions: the challenges and opportunities for Open Science in a historical context, local understanding and definitions of open and citizen science, motivation for participation, balancing outcomes in a grassroots citizen science project, as well as community mobilization. Finally, the conclusions highlight the way forward and the lessons learned during the implementation of this experimental project in Kyrgyzstan.

Case Study: The Kyrgyz Mountains Environmental Education and Citizen Science Project

Kyrgyzstan's Naryn province, where this project is being implemented, is characterized by remoteness, livestock-based livelihoods, low infrastructure development, as well as the highest poverty incidence in the country. In Soviet times, scientists began raising concerns about land degradation, mainly linked to overuse of natural resources (Kerven et al. 2012). After the collapse of the Soviet Union, the situation worsened, accompanied by a deterioration in the transportation infrastructure, water supply, and public buildings, as well as a drastic reduction of funds for research and monitoring of the environment. The new republics inherited far more research institutes and scientists than their economies could support. However, However, many of them have not been mentioned, leading to underfunded, ill-equipped, overstaffed, and ineffective institutes, with qualified staff moving on to better-paid and more relevant jobs. At the same time, data-sharing and linkages between research, education, and local-level policy have decreased, leaving remote areas with a lack of information for local environmental decision making. A closer collaboration with communities, generating a better understanding of their local environments, would be a very valuable contribution, especially related to education, in Kyrgyzstan.

Although environmental analyses abound for Central Asia in general, there is almost no data available at the local level or that differentiates between ecosystems and altitude levels within the highly diverse Central Asian ecological landscape. Additionally, in order to confront a poor understanding of environmental challenges and limited awareness of opportunities for change, it is instrumental to introduce locally embedded environmental education for the younger generation, who will prove primarily responsible for coping with and adapting to a rapidly changing environment (Gareeva and Maselli 2008; Schuler, Dessemonter, and Torgashova 2004; Mestre, Ibraimova, and Ajibekov 2013; UNDP 2006). During the World Summit on Sustainable Development held in Johannesburg in 2002, the idea was expressed that a lack of education and a low level of knowledge within the population on issues of sustainable development are possible reasons for existing problems in the environmental, social, and economic spheres (CAREC 2007).

While analyses on climate change conclude that Central Asia is exposed to one of the highest rates of adverse effects of climate change (Bizikova et al. 2011), additional challenges have arisen due to decades of mismanagement of natural resources. This includes the overgrazing of pastures, inefficient water and energy management, degradation of soils due to unsustainable agricultural practices, uncontrolled mining, loss of biodiversity, and increasing conflicts over natural resources (Gareeva and Maselli 2008; Schuler, Dessemonter, and Torgashova 2004; Mestre, Ibraimova, and Ajibekov 2013; UNDP 2006). Over half of Kyrgyzstan's GDP is derived from climate-sensitive and water-dependent activities, making the country highly vulnerable to the adverse impacts of climate change, and, in particular, decreased water supply, increased frequency and intensity of extreme weather events, and threats to ecosystems, livelihoods, and the health of the local populations (World Bank 2011). Understanding and observing these dynamics is therefore instrumental to supporting Kyrgyzstan's adaptation strategies (Buytaert et al. 2014).

The KMEECS project, implemented jointly by the nongovernmental organization (NGO) CAMP Alatoo and the University of Central Asia, started in 2015 and applies a transdisciplinary approach to knowledge generation. It combines citizen science at the community level, environmental research, and teacher training to foster awareness of and interaction with the local environment. At the same time, it aims at generating locally relevant data on the environment in the mountains of Kyrgyzstan. The project pilots the introduction of low-cost environmental field courses on water monitoring in schools in the mountain communities of Kyrgyzstan's Naryn province. Based on a citizen science approach, students analyze and generate data on their local environment to foster understanding of the changing environmental dynamics of local water resources. This project generates local-level data for understanding the changing environmental dynamics through water resources monitoring.

In a participatory curriculum development process, science teachers and students from ten schools in Ak-Talaa, Naryn, and At-Bashy districts contributed their knowledge, experience, and ideas for creating a citizen science curriculum on water monitoring. At the same time, local scientists from national-level academia were invited to contribute to the definition of meaningful parameters to be monitored in order to make measurements useful beyond the local level. As such, this project combines different development and research goals, stakeholders, and levels of intervention, which have proven to be partly contradictory during research implementation as will be described in the next section.

The project also analyzes the stakeholders involved in implementing the project. A multi-stakeholder participatory process for developing and testing a citizen science-based teaching manual for schools in rural areas of Naryn involves several degrees of participation (Arnstein 1969) and different degrees of activity and passivity within this process (Pretty 1995) for different stakeholders. This analysis generates insights on how OCS principles are applied and governed in a multi-level and multi-stakeholder process, with the aim of creating localized environmental education resources for remote schools in Kyrgyzstan.

Figure 12.1. Locating the KMEECS Project Villages Within a Regional and Global Context

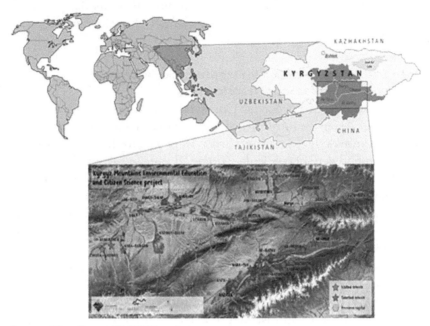

Source: Aline Rosset/KMEECS project.

Participatory Action Research was used as the main methodological framework for this project. The activities conducted within the project include motivation-based selection of science teachers (chemistry, biology, geography, and physics), round-table meetings for participatory curriculum development, situational analyses in the selected villages, joint field visits, training of teachers and students on water resource monitoring, workshops on participatory mapping and open hardware, as well as a public exhibition at the intersection between art and science, presenting locally relevant research results from every school at the national and local level.

Findings: Citizen Science on a Grassroots Level

Although new to most stakeholders and partners involved in the project, Open Science and citizen science coupled with environmental education have so far received much positive feedback and interest in Kyrgyzstan. At the same time, the project team also encountered many challenges during implementation. This section summarizes the perception of different stakeholders regarding the potential of citizen science in combination with environmental education in a remote mountain context.

Open Science and Education in Kyrgyzstan: Soviet Legacy and Opportunities for Social Transformation

The breakdown of the Soviet Union impacted the development of all Central Asian countries. In the case of education, research, and access to knowledge, a shift occurred from a centralized educational policy, unified school curricula, widely dispersed research networks, and large financial flows that were equally allocated by the central government to even the remotest regions to independent educational systems with significant challenges. School enrolment rates and the quality of education in Kyrgyzstan have regressed considerably since the late 1990s; this is particularly dramatic in remote areas. Declining quality has resulted mainly from budgetary neglect, which led to depleted stocks of textbooks and other educational materials, underpaid, under trained, and overburdened teachers, and the deterioration of school infrastructure (Mertaugh 2004; UNICEF 2008). Curricula that are overly theoretical allow hardly any scope for students to learn through practical and locally adapted teaching methods (UNICEF 2008).

The KMEECS project is one of few initiatives utilizing OCS in Kyrgyzstan. This is not due to a lack of ideas, but rather to a long-lasting culture of restricted information flow, mistrust, and bureaucratic regulations that originated in the Soviet era, inhibiting the deployment of a culture of openness in Kyrgyzstan. While this enduring legacy has largely shaped the complex and rigid political hierarchies of present-day Kyrgyzstan, initiatives on openness are burgeoning in the country, which is arguably the most open and democratic of the five Central Asian republics (Schenkkan 2015). Just to mention a few of these initiatives, there is a large open data movement, lobbying for the public availability of government data; the first hackathons have taken place; and there have been several events and initiatives highlighting the benefits of openness for business, democracy, and citizen information. The University of Central Asia also implements another citizen science case study. The goal is to involve community stakeholders in data-driven environmental decision making through Environmental Virtual Observatories (EVOs) in mountain areas. Another initiative is the "coalition for open education in Kyrgyzstan," which coordinates topics like the creation and dissemination of open resources for education, concentrating on fostering citizen contributions, e.g., Kyrgyz language, Wikipedia, OpenStreetMap, as well as policy dialogue on copyright. The latter is a particularly sensitive topic as there is still much confusion around the term "openness". Some perceive openness as a mere listing of resources or library catalogues on the Internet, without granting access to the general public. Openness is often understood as being equal to the use of ICTs and an increased online presence. Nevertheless, the interest in open approaches is increasing across all sectors in Kyrgyzstan as reflected in the *Global Open Policy Report 2016* (Wiens and Tarkowski 2016). In the report, Kyrgyzstan ranked among the ten most advanced countries in promoting open policy across four sectors: education, science, data, and heritage. In January 2017, Kyrgyzstan's parliament adopted amendments to the Intellectual Property Rights Law, which mandate that all publicly funded resources in all spheres must be made publicly available.

As for citizen science, the persistence of a "Soviet mindset" also challenges its uptake as a community-endorsed concept with legitimate scientific value. While citizen science has the potential to overcome entrenched legacies by empowering communities to engage in the creation and production of "their own" relevant knowledge, this participatory approach to science faces challenges due to many people in transitional post-Soviet contexts still tending to rely on external expertise rather than developing their own capacity (Buytaert et al. 2014). Some beneficiaries of the current social status quo—often former elites—also have an interest in preserving their status and privileges, leading to asymmetric power relations and a lack of trust in local and governmental decision-making institutions. Yet, in Kyrgyzstan, as a quickly democratizing republic, the involvement of local people in governing processes is steadily increasing, particularly at the municipal level.

The teachers and students involved in the KMEECS project jointly decided that they would like to focus their research on water, due to the crucial importance of water resources for their villages. One of the reasons for this choice was that the government does not have the capacity to conduct water monitoring at the local level. This led to high expectations among teachers and students that they would be able to fill this gap, particularly related to water quality measurements. Some analyses are, however, too complex to be conducted outside professional laboratories (and some cannot be conducted in Kyrgyzstan at all), although the results would be important for most communities (e.g., heavy metals in areas close to mining sites and bacterial contamination in all villages due to a lack of water treatment and uncontrolled infiltration of manure and leakage of household sewage into the drinking water system).

At the same time, the teachers found out that there are many differences in water resources in the ten villages. Some have abundant water resources, while others lack drinking water or water for irrigation. Some have access to clean drinking water, while others face significant challenges concerning water quality and health, even mentioning the presence of unidentified worms during summer months. At present, there is little opportunity or local capacity in rural areas of Naryn to monitor water resources. This was also confirmed by our research, as more than half of the interviewed teachers highlighted that community institutions and individuals approached them with requests to analyze their water and provide recommendations to the community. "Most people in our village are losing trust in the piped water we used for decades. It is not being treated; we even found some cadavers of livestock in our reservoir, and many feel that our health is getting worse. So almost everyone started digging groundwater wells in their backyards. We received a large number number of requests to conduct analyses of people's backyard water, although we mentioned that we cannot test for all potential threats," said a chemistry teacher from the Naryn district.

Interestingly, even though governmental institutes conduct more accurate analyses, that include more parameters, in some communities, people do not trust their results. "Sometimes when we send water samples to governmental laboratories, we even think that they mix the results intentionally to hide sensitive information. Our analyses are much simpler, but at least we are sure that they are not manipulated. And based on that, the villagers start thinking and draw their own conclusions on what could be improved in our water management," said a physics teacher from the At-Bashy district. Finding an inclusive research approach that reconciles local environmental knowledge and modern scientific approaches that generates robust monitoring results and trust at the local level appears to be a highly important factor for challenging the top-down, "expert"-oriented scientific legacy that remains in Kyrgyzstan.

Understanding Open Science: Perspectives from the Local Level

Open Science is a new concept for Central Asia, and it is still very difficult to grasp. Nonetheless, the stakeholders involved are enthusiastic and highly motivated to advocate for OCS. This is particularly true for teachers and students, our main partners in the ten villages, who showed an unexpected amount of motivation and involvement. Although the involvement of students in discussions on the development of learning resources was a novelty for all participants, the young citizen scientists (in particular, girls) proved to be very active and interested during project meetings and events.

At the same time, it is still very unusual for participants from remote villages to be involved in decision-making processes, particularly concerning scientific approaches. This has been visible at several points in the process of developing monitoring experiments, tools, and activities for the project. While teachers and students are very good at giving their opinion, attending workshops, participating actively, presenting results of their discussions, and giving feedback, it is very difficult to trigger concrete contributions to the content and to engage them in shaping the activities according to their experiences and wishes.

Particularly for Kyrgyz, as a language and culture background where Open Science and citizen science are not yet established, there is a need for a proper definition of the Open Science terminology, and how to meaningfully translate and explain it in local languages. There needs to be a discussion on what Open and Collaborative Science means at the local level, how the terminology is described and understood at different levels (policy makers, development organizations, rural stakeholders, etc.), and how it should be translated in order to be understood correctly. As in the rest of the Soviet Union, the Russian language dominated in Kyrgyzstan for over seventy years. The use of the Kyrgyz language was almost absent in education, science, public service, and commerce (Linn et al. 2005). With a decreasing command of Russian (in rural areas in particular) and a school system predominantly transferred to Kyrgyz (as the national language of the country), large parts of the scientific knowledge, still mainly generated in Russian for reaching a broader audience beyond Kyrgyzstan, become virtually inaccessible. Here, citizen science offers the opportunity to generate local knowledge and to present results in local languages. The issue of translation was also highlighted by a geography teacher from the Ak-Talaa district:

> It is great that all resources on Citizen Science and experiments are presented in Kyrgyz, so we can contribute and directly apply it to our lessons. However, it is apparent that some terms were directly translated from scientific Russian, as we don't have this terminology in Kyrgyz. It would be good to negotiate among us and agree on a practical, simplified translation that is understandable, rather than a literal adaptation that sounds very bulky.

Even if the terminology and the process of OCS is a new venture for project participants in Kyrgyzstan, many of them immediately connected to its principles and compared it to their own mindset:

> Even if I didn't tag it with a particular scientific approach so far, participatory science and citizen control over the development of our village are the backbone of my civic engagement. Citizen Science generates facts for the villagers; we can compare results over a certain time and make decisions based on them. This allows villagers to improve their scientific literacy for solving problems. For example, after mapping our village and the places where we get our drinking water, it became visible to anyone who wants to see it that the eastern part of the village is well provisioned with water points, while the rest isn't. Based on that, we can take action! (biology teacher from the At-Bashy district)

It became apparent that the effective application of citizen science in our case study partly depended on the identification of and collaboration with unique innovative individuals, or "openness champions," who were willing to consider and try new collaborative research, education, and communication approaches. Teachers also came up with their own definitions of citizen science, such as "scientific achievements of the village inhabitants themselves on topics that are interesting to and defined by them" (geography teacher, Naryn district); or "Citizen science can be conducted by ordinary people, based on simple methodologies. It is science that is no longer only for formal scientists. I found out that we and our students can equally qualify to conduct scientific investigations and might even have more knowledge than renowned researchers about the mechanisms in our local environment" (geography teacher, At-Bashy district). A geography teacher from the Ak-Talaa district also mentioned that citizen science reminds her of Russian geographers, visiting the region before the collapse of the Soviet Union, who always involved community members in mapping the area, "because locals know their surroundings better than others. But they were not regarded as scientists, rather as guides or informants."

Motivation: Why Should Teachers Participate in Citizen Science?

Motivation plays a key role for active participation in the project's activities. The sixteen teachers participating in the development and implementation of citizen science activities were selected based on the provision of written motivation letters, which is an unusual procedure. It has, however, been highly effective, ensuring ninety-eight percent attendance and active participation during the project workshops as well as unexpectedly high levels of enthusiasm and creativity. At each meeting, workshop, and event, there were at least two students per school in attendance. Students participated actively during meetings and workshops, particularly girls, on topics linked to chemistry, biology, geography, and physics, with equal attendance from boys and girls during the group meetings. Similarly, out of the sixteen teachers permanently involved in the project, eight are women and eight are men. Interestingly, this finding reflects global UNESCO statistics that have found that female researchers in Kyrgyzstan represent 49.4 percent of total researchers in the country (UNESCO 2018). It should be noted that Central Asian countries as a whole performed significantly better than Western countries in this regard, with 48.1 percent representation from female researchers, compared to 32.3 percent in Western Europe and North America (UNESCO 2018). Although more research is needed, the employment of locally relevant Open Science could be an important opportunity for making science more accessible for traditionally underrepresented groups, including women.

As far as teachers are concerned, we identified two different levels of motivation—one level being related to institutional motivation, in contrast to teachers' individual motivation. On one hand, some teachers seemed to be very focused on the toolbox and the material contributions they would receive for conducting the experiments. This can largely be attributed to the evaluation system based on which schools and teachers are monitored and rewarded: participation in projects must lead to material benefits for the schools. A Chemistry teacher from the Naryn district said:

> We need to show our director that this project will bring some visible benefit for our school; otherwise, he would not allow us to attend the project meetings. For him, this means equipment and tools. For us, the most important thing is to be involved and to learn new interesting approaches that connect learning to reality for our students.

Due to this constant pressure from school directors, teachers were repeatedly asking for the research equipment they would receive as a toolbox for environmental monitoring activities. A chronic lack of equipment for applied teaching and laboratory work also ranked high among teachers' individual motivation, although the materials provided were mainly low-cost and, wherever possible, do-it-yourself:

> When it comes to practical experiments in school, we are still applying Soviet methods, which are quite complex and don't make the link to real-life examples that children can grasp and understand. They also require specific tools, which are now mostly broken, and chemicals, which are too old and far over expiry date. What I mostly like about this approach is that it uses simple tools and makes us think that we can also build some instruments ourselves. To be honest, we are very used to receiving ready materials for specific purposes. Who would have thought that it is possible to build our own microscope?! (chemistry teacher, Ak-Talaa district)

At the same time almost all the teachers assured us that other factors equally determine their own motivation, particularly highlighting new ideas for interactive methodologies, outdoor education, and scientific curiosity. Thus one biology teacher from the At-Bashy district commented:

> The teaching approach was also completely different from what we would like to see in today's education. The teacher was seen as a sort of dictator during lessons, and the tools and experiments

would only be used by the teacher. This led to a limited creativity of teachers and students. Now we are really in need of interactive methodologies and ideas for bringing change and fostering inquiry.

Another biology teacher from the Ak-Talaa district said: "Open Science would be a very valuable approach to be included in the school program, particularly approaches fostering scientific outdoor investigations and instructions on how to construct and share our own research tools."

Most participants indicated they were interested in the research components related to their own discipline, as this is what they teach. However, many mentioned transdisciplinarity aspects as a great learning outcome for themselves and their students: "As a chemistry teacher, I have so far been focusing only on my own background. Here, we looked at water from different perspectives, including biological, geographical, chemical, and physical measurements. This was a turnkey experience for me, showing me that we should not stay within the limits of our own garden for understanding the whole picture," mentioned a chemistry teacher from the Naryn district.

Balancing Participation, Community Action, and Science Outcomes

One of the distinctive features of participatory research is its focus on issues of interest and concern to the participants themselves (Robottom and Sauvé 2003). However, in our case, this proved to be a factor hindering a stronger engagement of scientists from national-level research institutes in Kyrgyzstan. A common answer when discussing the potential of citizen science for tackling the lack of local-level environmental data in mountainous regions with scientists from national institutions is reflected in this reply from a hydrologist at a Kyrgyz governmental agency:

> There is no doubt that this project is nice from an educational point of view. It is great for the kids to get the opportunity to conduct hands-on analyses. However, I don't see much value for science or governmental institutions. These are kids after all; they cannot comply with scientific requirements, even more if you work in remote areas. Also, for conducting meaningful research, we need reliable tools, which are too expensive to be "wasted" on an educational project.

The following summarizes most of the concerns that we identified when talking to scientists and researchers from national and regional-level research institutes: 1) the age of the citizen scientists, and the fact that they are not trained scientists—children and youth are perceived as incapable of collecting rigorous data; 2) the geographic focus—remote mountain communities, lacking access to modern infrastructure, education, and knowledge—is even less legitimate for conducting scientific activities; 3) the adoption of a low-tech, low-cost approach—scientific measurements are only valuable if conducted with a standardized high-tech infrastructure, which is operated by specialists formed for this purpose; and 4) the collaborative process for defining research priorities. As such, this project identified a trade-off between encouraging grassroots participation in defining research topics and a demand for local data from academia and practitioners. This suggests a tension between citizens and traditional science, since giving more power to steer the research process to one of these main stakeholders would reduce the decision-making power of the other (unless the common goal of the research is to conduct participatory research together, with no prior expectations as to the outcomes). Among local-level stakeholders from rural schools and communities, the picture was much different. Teachers and students reacted similarly in the beginning, but, after the second meeting, they were confident that they were in fact equally experts when it comes to their local environment.

Moreover, the project raised much interest among international development and research organizations, educational institutions at all levels in Kyrgyzstan, as well as among local NGOs involved in environmental projects in rural areas. National institutions mainly highlighted the educational value, while organizations with linkages to global discourses on participatory approaches for sustainable natural resource management were equally interested in the opportunities to conduct environmental monitoring. This highlights an increasing interest in OCS and its benefits for connecting education, research, and development in Central Asia. Accordingly, funding for a follow-up project on phenology and climate science could be secured, with a stronger engagement of scientists and development practitioners interested in data outcomes. This will increase the visibility and usability of the outcomes, but—as mentioned in the trade-off above—at the same time, it will probably decrease the control of teachers and stakeholders at the local level.

As mentioned above, during workshops and field visits to the ten participating schools, it became apparent that each of the villages presents a different situation concerning water availability and quality. Consequently, each school wanted to highlight different research questions and communication tools for presenting their findings to the community. This is why it appeared necessary to increase the focus on individual cases for scientific investigation. While all the schools conducted the same basic analyses for monitoring water in their community, each school is also focused on a specific challenge for their village, investigating this issue and visualizing it in a tangible manner. These scientific projects will be part of a competition between schools and will appear in a travelling exhibition, visiting each village in the region of Naryn and Bishkek, the capital of Kyrgyzstan. Such an exhibition was planned from the beginning of the project, but, in order to highlight the diversity of water challenges, it is now receiving more attention than initially planned.

This participatory definition of local research priorities generated a wealth of individually relevant environmental information that can be used to foster civic action at the local level. A public exhibition also presents the achievements of the involved students and teachers, showcasing their scientific investigations of local environmental challenges at local, provincial, and national levels. However, there still remains a challenge that comparable environmental data are needed at a larger scale; for this, generalizable parameters and indicators that are relevant for science and policy need to be defined with the help of academic and local scientists. As the head of a local environmental NGO indicated:

> People are making policy decisions based on a lack of good baseline data, but there are no financial and human capacities to conduct relevant measurements across a large geographic area. The involvement of students could well combine the generation of data along with environmental education goals. But this means that the value of this work needs to be recognized, and, for this, you need support from scientists and practitioners who are ready to consider this information for decision making—and therefore also show willingness to contribute to the definition of indicators to monitor.

Community Mobilization and Citizen Science-based Activism

If Open Science is a new concept for teachers in Kyrgyzstan, then sharing knowledge among peers has certainly been broadly practised for many years. Teachers mentioned different channels that they use to plan and disseminate their own ideas and the citizen science methodologies jointly developed. Among existing platforms for sharing experiences, they mentioned regional teachers' workshops and school academic competitions for sharing beyond their community. As tools for sharing their experiences within their school and community, they listed extracurricular student working groups, school newspapers, collaboration with other teachers to introduce transdisciplinary approaches, open lessons, and community meetings involving parents and other villagers.

In two villages, students and teachers organized community events to clean riverbanks, at the same time presenting information on water monitoring and the importance of keeping waste, livestock, agro-chemicals, etc., away from waterways. The fact that the tools developed are simple and low-cost and that they focus on the use of locally available materials was highlighted as a great advantage for sharing: "It was great to develop experiments with readily available materials, so that they are replicable also with limited funds. This way we will be able to share our manual with other schools that have not been part of the project so far, if it will be possible to print many copies and disseminate them broadly," mentioned a biology teacher in the At-Bashy district.

Some teachers and students have already initiated cooperation with local decision makers to begin a dialogue on tackling their communities' challenges regarding water. This is particularly the case where water is not equally available for all inhabitants or where there are concerns about water quality. As one of the ways to disseminate results, a geography teacher from At-Bashy indicated the value of face-to-face communication: "After monitoring our water resources, students go home and inform their parents about what they found out. This is how information is moving around here, and then questions and requests for clarification come back to us, sometimes through students, sometimes directly through their parents."

Based on the high diversity of water investigations conducted by the ten schools, an exhibition of citizen science projects was identified as an experimental way to distribute information and reach out to different audiences at various levels. Science and art naturally overlap. Artists and scientists both study their environment and learn to transform information into something else. Based on this, the project participants created visual outcomes displaying local research questions and results in the framework of an exhibition and competition.

Conclusions

> Many opportunities for community citizen science occur locally, through development of practices that match the investigation of resources with the needs of their users. Substantial challenges remain at larger scales...The knowledge needed to inform action requires an interdisciplinary science that draws on the observations, skills, and creativity of a wide range of natural and social scientists, practitioners, and civil society. (Chapin et al. 2011)

Similarly, our case study in Kyrgyzstan highlights that a focus on particular research projects at the local level implies challenges for significant application on a larger scale. Also, Open Science—and, in particular, citizen science conducted by children—is not yet understood as serious science by the national academia in Central Asia. Therefore, the potential contribution of citizen science to environmental monitoring and education needs to be better understood and advocated in Kyrgyzstan. This project is a contribution toward testing the implementation of hands-on, outdoor activities for schools, requiring very little equipment, and demonstrating the transdisciplinarity of environmental challenges. Through an exhibition of citizen science projects conducted in schools and a policy dialogue, it is intended to provide input to integrate inquiry-based approaches when elaborating new standards for education in Kyrgyzstan.

Returning to the applicability of Open Science for social transformation in Kyrgyzstan and the balance between educational, scientific, societal, and policy goals, this case study has demonstrated clear benefits on the educational level and has also contributed to local-level public engagement in societal discourse around water management. At the same time, large-scale scientific outcomes and policy goals have not been the main focus of the research. However, as reflected in the rapid developments of open policy in Kyrgyzstan mentioned earlier, it will be very important to guarantee continued intersectoral and multi-level coordination between stakeholders to ensure that small initiatives can be embedded in a policy-level dialogue for fostering an inclusive culture of openness in Kyrgyzstan.

In order to allow further dissemination of citizen science coupled with environmental education in Kyrgyzstan, it will be instrumental to spark the interest of scientists and practitioners in contributing to define research questions and indicators to be monitored on a broader scale—even if this reduces the possibility of adapting the research process to individual community needs. A clear scientific interest in using environmental data beyond the local level can also act as a motivating factor for schools and local activists to collect information over longer periods of time. The motivation of teachers and students has so far been high, and partner organizations interested in piloting research with schools on other environmental topics have been found.

This case study can be seen as a pilot project, testing the implementation of several concepts related to Open Science on a small scale in a rural area of Kyrgyzstan. Citizen science and citizen participation in research, the potential of open hardware, projects such as OpenStreetMap, as well as Open Access to information, are being discussed in networks, think tanks, and meetings, and an increased interest in these approaches is already visible. However, a small project operating on the local level is not enough to spark a culture of openness. A major prerequisite to rooting a culture of openness in Kyrgyzstan will be to raise awareness about the benefits of public participation in scientific research and open information, coupled with a clarification and adaptation of laws concerning access to, dissemination of, and creation of information.

At the moment, Kyrgyzstan and Central Asia are on the brink of plunging into the digital age of information, with an awakening culture of openness concentrated in urban centres. This also leads to growing inequalities, as the rural areas still lag behind these developments in terms of a culture of openness as well as availability of technology, connectivity, and education. By involving rural stakeholders, this project contributes to addressing the gap that exists between rural and urban areas and to giving a voice to people who have not yet been involved in developing ideas and showcases for education. The deployment of OCS—as well as open and collaborative education, data, information, etc.—can greatly contribute to social transformation by reducing the gaps not only between government agencies and civil society, but also between rural and urban areas in Kyrgyzstan, if care is taken to make the openness movement inclusive and not to replicate existing hierarchies.

Notes

1. While the two biggest cities of Kyrgyzstan (Bishkek and Osh) are located at an altitude of 800 and 960 m.a.s.l., respectively, eighty-seven percent of Kyrgyzstan's total territory lies at altitudes of 1,500 m.a.s.l. and higher, and more than forty percent of the whole territory lies above 3,000 m.a.s.l. (UNDP 2002). Therefore, disparities between centre and periphery overlap with disparities between low-lands and high mountains.

References

Arnstein, Sherry. R. 1969. "A Ladder of Citizen Participation." *Journal of the American Institute of Planners* 35 (4): 216–24.

Bizikova, Livia, Hilary Hove, Jo-Ellen Parry, and International Institute for Sustainable Development. 2011. "Review of Current and Planned Adaptation Action: Central Asia."*Adaptation Partnership*. www.adaptation partnership.org.

Bonney, Rick, Jennifer L. Shirk, Tina B. Phillips, Andrea Wiggins, Heidi L. Ballard, Abraham J. Miller-Rushing, and Julia K. Parrish. 2014. "Next Steps for Citizen Science." *Science* 343 (6178): 1436–37. https://doi.org/10.1126 /science.1251554.

Buytaert, Wouter et al. 2014. "Citizen Science in Hydrology and Water Resources: Opportunities for Knowledge Generation, Ecosystem Service Management, and Sustainable Development." *Frontiers in Earth Science* 2 (26): 1–21.

CAREC. 2007. *Progress Review on Education for Sustainable Development in Central Asia: Achievements, Good Practices and Proposals for the Future*. Almaty: Central Asia Regional Environmental Center.

Chapin, F. Stuart, Mary E. Power, Steward T. A. Pickett, Amy Freitag, Julie A. Reynolds, Robert B. Jackson, David M. Lodge, Clifford Duke, Scott L. Collins, Alison G. Power, and Ann Bartuska. 2011. "Earth Stewardship: Science for Action to Sustain the Human-Earth System." *Ecosphere* 2 (8): 1–20. doi:10.1890/ES11-00166.1

Cohn, Jeffrey P. 2008. "Citizen Science: Can Volunteers Do Real Research?" *BioScience* 58 (3): 192–7.

Dickinson, Janis L., Jennifer Shirk, David Bonter, Rick Bonney, Rhiannon L. Crain, Jason Martin, Tina Phillips, and Karen Purcell. 2012. "The Current State of Citizen Science as a Tool for Ecological Research and Public Engagement." *Frontiers in Ecology and the Environment* 10: 291–7.

Gareeva, Aida, and Daniel Maselli. 2008. *Natural Resource Management for Sustainable Livelihoods. Challenges and Trends in Central-Asian Mountain Regions*. Bishkek: CAMP Alatoo.

Gommermann, Luke, and Martha C. Monroe. 2012. "Lessons Learned from Evaluations of Citizen Science Programs." *EDIS* Publication FOR291. Gainesville, FL: University of Florida. http://edis.ifas.ufl.edu.

Haklay, Muki. 2012. "Citizen Science and Volunteered Geographic Information – Overview and Typology of Participation." In *Crowdsourcing Geographic Knowledge: Volunteered Geographic Information (VGI) in Theory and Practice*, edited by Daniel Z. Sui, Sarah Elwood, and Michael F. Goodchild, 105–22. Berlin: Springer.

Kerven, Carol, Bernd Steimann, Chad Dear, and Laurie Ashley. 2012. "Researching the Future of Pastoralism in Central Asia's Mountains: Examining Development Orthodoxies." *Mountain Research and Development* 32: 368–77. https://doi.org/10.1659/MRD-JOURNAL-D-12-00035.1.

Linn, Johannes et al. 2005. *Central Asia Human Development Report: Bringing Down Barriers; Regional Cooperation for Human Development and Human Security*. New York, NY: UNDP.

Mackechnie, Colin, Lindsay Maskell, Lisa Norton, and David Roy. 2011. "The Role of 'Big Society' in Monitoring the State of the Natural Environment." *Journal of Environmental Monitoring* 13: 2687–91.

Mertaugh, Michael. 2004. "Education in Central Asia with Particular Reference to the Kyrgyz Republic." In *The Challenge of Education in Central Asia*, edited by Stephen P. Heyneman, and Alan J. DeYoung, 153–80. Greenwich, CT: Information Age Publishing.

Mestre, Irène, Aliya Ibraimova, and Bilimbek Ajibekov. 2013. *Study of Conflicts over Pasture Resources in the Kyrgyz Republic*. Bishkek: CAMP Alatoo.

Nov, Oded, Ofer Arazy, and David Anderson. 2011. *Dusting for Science: Motivation and Participation of Digital Citizen Science Volunteers*. Seattle: iConference.

OECD. 2015. "Making Open Science a Reality." *OECD Science, Technology and Industry Policy Papers* 25. Paris: Organisation for Economic Co-operation and Development Publishing.

Pretty, Jules. 1995. "Participatory Learning for Sustainable Agriculture." *World Development* 23 (8): 1247–63.

Riesch, Hauke, Clive Potter, and Linda Davies. 2013. "Combining Citizen Science and Public Engagement: The Open Air Laboratories Programme." *Journal of Science Communication* 12 (3). https://doi.org/10.22323/2.12030203.

Robottom, Ian, and Lucie Sauvé. 2003. "Reflecting on Participatory Research in Environmental Education: Some Issues for Methodology." *Canadian Journal of Environmental Education* 8: 111–28.

Roy, H.E., M. J. O. Pocock, C. D. Preston, D. B. Roy, J. Savage, J. C. Tweddle, and L. D. Robinson. 2012. "Understanding Citizen Science &

Environmental Monitoring." Final Report on behalf of UK-EOF. *NERC Centre for Ecology & Hydrology and Natural History Museum.*

Schenkkan, Nate. 2015. "Central Asia's Island of Democracy Is Sinking," *Foreign Policy News.* https://foreignpolicy.com/2015/09/30/central-asias -island-of-democracy-is-sinking-kyrgyzstan/.

Schuler, Martin, Pierre Dessemonter, and Liudmila Torgashova. 2004. *Mountain Atlas of Kyrgyzstan,* NSC Bishkek, EPF Lausanne.

Silvertown, Jonathan. 2009. "A New Dawn for Citizen Science." *Trends in Ecology & Evolution* 24: 467–71.

Shirk, Jennifer L., Heidi L. Ballard, Candie C. Wilderman, Tina Phillips, Andrea Wiggins, Rebecca Jordan, Ellen McCallie, Matthew Minarchek, Bruce V. Lewenstein, Marianne E. Krasny, and Rick Bonney. 2012. "Public Participation in Scientific Research: A Framework for Deliberate Design." *Ecology and Society* 17 (2): 29. http://dx.doi.org/10.5751 /ES-04705-170229.

UK-EOF. 2011. "Citizen Science Observations and Monitoring: Scoping Requirements, Knowledge Exchange and Finding Potential Synergies." *UK Environmental Observation Framework.*

UNDP. 2002. "Human Development in Mountain Regions of Kyrgyzstan." *National Human Development Report.* Bishkek: United Nations Development Programme.

———. 2006. *Kyrgyzstan: Environment and Natural Resources for Sustainable Development.* Bishkek: United Nations Development Programme.

UNESCO. 2018. *Women in Science.* Fact Sheet No. 51. http://uis.unesco.org /sites/default/files/documents/fs51-women-in-science-2018-en.pdf.

UNICEF. 2008. *Education in Kyrgyzstan.* Country profile. http://www.unicef .org/ceecis/Kyrgyzstan.pdf.

Wiens, Kelsey, and Alek Tarkowski. 2016. *Global Open Policy Report 2016.* Open Policy Network.

World Bank. 2011. *World Development Report 2011: Conflict, Security, and Development.* Washington, DC: World Bank.

CHAPTER 13

Open Science and Social Change: A Case Study in Brazil

Sarita Albagli, Henrique Parra,
Felipe Fonseca, and Maria Lucia Maciel

Abstract

The community of Ubatuba, in São Paulo, Brazil, is located in a dense rainforest region. A diverse mix of Indigenous communities, researchers, activists, and policy makers are interested in the area. Thus, it makes a compelling case study for examining the potential of Open and Collaborative Science (OCS) from a sustainable development perspective. This project draws on a reflective, action-based research approach to understanding the institutional, cultural, and political challenges involved in the adoption of an OCS approach for development in Ubatuba, Brazil, by interacting with a variety of different actors. The authors conclude that, on one hand, OCS does create new spaces and methods for traditionally marginalized groups to engage in scientific discussions and local problem-solving, mainly in controversial and conflict situations and as a condition for resilience and political struggle for alternative paths of development. On the other hand, the very idea of openness is under dispute: What (Open) Science and for whom?

Introduction*

Open Science movements[1] have gained traction worldwide, most recently in so-called emergent and developing countries, or the Global South (Albagl, Maciel, and Abdo 2015). However, the debate on science and technology's role in social development is not new. It has taken place within Southern contexts, particularly in Latin American countries, since the 1960s (see Sunkel and Paz 1970; Herrera 1972; Varsavsky 1972; Morel 1979; and Fals-Borda 1981, among others). A novelty in this field is the growing dissemination of experiences that value the adoption of methods and practices in open and collaborative knowledge production, taking advantage of the opportunities opened up by information and communication technologies. In particular, such methods are conveyed as enabling the necessary conditions for more sustainable and participatory development strategies.

At the same time, these debates must be situated within a global context of traditional scientific knowledge production that is inherently exclusive unequal, and inaccessible to the majority of human beings. There is thus a need for development alternatives—or alternatives to development—that can mobilize opportunities for science to experiment with more open and collaborative approaches to knowledge production. However, since this is an emergent discourse, particularly in Brazil, there is a scarcity of knowledge about how these ways of working can be applied in practice, including their potentialities, obstacles, and requirements.

This chapter aims to contribute to filling this gap by presenting results of a case study on the possibilities and limits of Open and Collaborative Science (OCS) in social change processes, based on the results of an action-research project developed in the Ubatuba municipality, on the northern coast of the State of São Paulo, Brazil.[2] The question underlying our investigation was to what extent Open Science may improve forms of co-production of knowledge between academia and other social groups, and hence contribute toward improved conditions for vulnerable actors to influence development strategies. From the outset, various questions arise: What development? What science? What openness and collaboration?

The methodology was organized along two axes—practical learning and critical research—developed through the following actions:

(1) promotion of open workshops, working groups, seminars, and mentoring activities—stimulating discussions about and experimentation with Open Science practices and tools with local actors,[3] including civil society and government agents, high school and elementary school students, open knowledge and free digital culture advocates, as well as scientific

research groups; focusing on awareness, training in specific tools and methodologies for possible uses of OCS in local development issues, and inviting practitioners to share their own developments; monitoring and analysis of these activities and results, including short interviews and questionnaires with participants;

(2) participant observation of public meetings and activities, selected for their potential relationship with the open and collaborative production of knowledge pertaining to local development issues;

(3) data collection for socio-economic characterization of the Ubatuba municipality and its major development challenges;

(4) interviews with key actors—local government, scientific researchers, non-governmental organizations (NGOs), and protected areas managers—asking about their views on local development strategies and also about their needs, interests, and resistance toward the adoption of OCS values and practices; and

(5) social networks analysis to better understand local interactions among actors—and also a characterization of scientific publications related to Ubatuba as a field or object of study (authorship, institutional, and field of research distribution; open and closed access to papers).

A necessary component of the research methodology was the development of communication channels and documentation strategies, including a project wiki,[4] website/blog,[5] mailing lists, community radio programs,[6] and videos.[7]

This chapter is organized into four parts: The first part situates the research in the area where it was conducted, which comprises the empirical and territorial framework for analysis. This is followed by a presentation of the main findings and conclusions obtained from the practical experiments with Open Science, developed in partnership with local actors, and the derived analysis from it. The penultimate part of the chapter discusses the institutional dimension of mobilizing Open Science in social change processes. And finally, we share some concluding remarks to indicate limits and challenges faced by Open Science practices toward alternative development.

Ubatuba: Development in Dispute

The Brazilian municipality of Ubatuba is located on the northern coast of the State of São Paulo and occupies an area of 723,883 km². In 2016, it had an estimated population of about eighty-seven thousand inhabitants (according to the 2010 Brazilian population census, 59.2 percent of these identified as white, while 39.4 percent identified as mixed race and black, 1 percent identified as Asian, and 0.4 percent as Indigenous (Instituto Polis 2013). Ubatuba is part of the Atlantic Forest region, a strategic and vulnerable environmental area, which comprises various ecosystems with a high level of endangered biodiversity. A significant number of traditional communities (Indigenous, *quilombolas*,[8] and *caiçaras*[9]) live in this region, and they face greater difficulty in having their basic rights recognized and in having access to goods and services available to other segments of the population, which makes them more vulnerable to social exclusion. In the Ubatuba region, 86 percent of the territory lies within the Serra do Mar State Park. This region is also the home of the first protected marine area in Brazil.

Ubatuba: Northern Coast of São Paulo, Brazil

Figure 13.1. Ubatuba municipality on the north coast of São Paulo, Brazil

Source: Wikimedia.

The region around Ubatuba is characterized by diverse and conflicting interests and development perspectives. These interests include: (1) the indirect impact of the activities of large oil companies (due to the emerging industry of "pre-salt" oil exploration[10] along the Brazilian coastline); (2) predatory tourism—that is, tourism that causes numerous negative social, environmental, and economic impacts, such as high pressure on the provision of sewage, garbage collection, and water supply infrastructure, besides high seasonality, low wages, and little commitment to local production— led by an aggressive construction industry focused on intensive real-estate development; and (3) the lifestyles of traditional local communities, living in the last forty years along the borders of the protected area.[11] Concurrently, there are some "alternative" local economic initiatives taking place, including the solidarity economy, family farming, traditional fishing, permaculture, and community-based tourism.

Going further, the Ubatuba Municipality is also a focus of significant scientific research activity in diverse fields, including biological and environmental sciences (particularly in plant and marine biology), oceanography and social sciences, and ethnography, and the humanities. Despite the abundance of scientific and academic research on and within the area, the conditions for local populations to access this knowledge and its socio-economic benefits are not guaranteed and remain an object of dispute. Scientific work could thus benefit local development initiatives and demands through more open and collaborative practices, contributing to increased public visibility and citizen participation, as well as facilitating closer linkages with a wide spectrum of local actors.

Ubatuba also congregates a rich body of knowledge produced by local and traditional communities, which is relevant to sustainable development strategies but is not sufficiently recognized and valued. Thus, conflicting demands and unequal power relations largely define the local political climate of the area. In general, local actors have unequal access to information and little influence on decision-making processes that directly affect their socio-economic well-being. This problem was very evident during the discussions about the revision of the Ecological-Economic Zoning (EEZ) of that region, a process that was followed closely by our project team. Marcos Tupã, coordinator of the Guarani Yvyrupa Commission (CGY1), representing the Guaraní Mbya Indigenous people in the south and southeast of the coastal region, expressed the lack of information and consultation of local populations in this way:

> We, traditional communities [*quilombolas*, *caiçaras*, Indigenous], have never been heard, invited, or consulted to discuss the zoning of our coastlines. We have Indigenous territories there in Silveiras [São Sebastião], Renascer [Ubatuba], and Boa Vista village, in Prumirim [Ubatuba], and all these processes are running over the guaranteed rights that are in the Constitution. We want to have space to be part of this mapping process and express our concern about the territory.

Notwithstanding, Ubatuba remains a site of diverse social, political, and economic experiments developed by local organizations, communities, and activists advocating for political empowerment and increased participation. Most of these actors have been shown to be sensitive to and mindful of the value of open and collaborative forms of knowledge production.

Open Science in Action

The activities we developed as part of the action-research methodology allowed us to foster debate on OCS and its potential usefulness toward development goals, while promoting dialogue among local actors. Our interactions with key actors and consequent development of a variety of core communication channels also worked to disseminate and reiterate a culture of Open Science and knowledge-sharing among diverse interest groups. At the same time, these actions facilitated capacity building around the use of alternative mechanisms for effective communication as well as highlighting the importance of citizen participation in knowledge production.

Research developed through interviews and participant observation helped us understand the perceptions of local actors about science and, more specifically, about Open Science. Although the term "Open Science" was not directly used by local groups, collaborative and transparent forms of information production and circulation constitute a recurring and strategic theme for agents involved in initiatives for local (and alternative) development, particularly in social and environmental management and policies. During our research, we discerned different meanings and perceptions of Open Science, and this was expressed in the involvement of different groups, in different ways, throughout the research.

Most groups consider access to scientific and technical information and knowledge as imperative for their success and stressed the importance—but also the obstacles and difficulties—of opening up scientific data about Ubatuba. As a technician of the North Coast Watershed Committee said in an interview for the project on October 24, 2016:

> Open Science serves the collective. Stored information is not information... . We often need information that we know was produced by a researcher, but we cannot access it to produce a monitoring report. ...There is a fear in the academy of information theft. It is a problem of taking possession of information.

It was also pointed out that (co)production of knowledge was important among different cognitive actors with diverse epistemic recognition and legitimacy, with an emphasis on intra- and extra-scientific collaboration and promoting dialogue between scientists and other social groups. Openness is also perceived as contributing to the development of critical consciousness on the unequal distribution of economic and environmental resources; for example, a coordinator of a regional research institute said, "I understand Open Science thus: making information available to generate more critical thoughts, people being empowered at least locally."

The project results confirmed our initial perception that social groups with a lower socio-economic status may benefit more from openness in terms of political empowerment. Openness, when defined as the democratization of access to and production of knowledge and information, proved to be a key factor for the promotion of citizen resilience within political struggle, improving the quality of participation, particularly in controversial situations regarding alternative paradigms for local development. It assists in revealing asymmetries and allows for subordinate positions to have more visibility and influence in decision-making processes.

This was quite evident throughout our monitoring and registering of public consultation and debates on the process of reviewing the EEZ[12] of the north coast of São Paulo, more specifically in the region of Ubatuba. The choice of the EEZ review process for empirical analysis has provided a more situated perspective on the relations between, on the one hand, production of and access to information on the territory and, on the other hand, the dispute over development models and territorial rights in general. This process demonstrated that local actors lack qualified information. Furthermore, these public hearings are generally addressed to those who possess "reasonable technical background," constituting an obstacle to wider social participation.

We also realized that the production and sharing of geospatial data constitute a demand of many actors and organizations to increase the quality of their intervention in territorial management. We thus decided to carry out an exercise in building a citizen science prototype to articulate sharing and visualizing of spatial data in an open and collaborative way. This experiment allowed: (1) exchange of experiences among different institutions, (2) collective construction of protocols on priority topics and (3) stimulation of synergy among future initiatives of common interest. This prototype indicated potential for increasing the quality of social participation in the local territorial planning processes.

Referring to the conflicts between different views of development for the region, one of the interviewees who works on a local social project expressed it this way: "Ubatuba for me is at the centre of the dispute of the counter-hegemonic forces in the North Coast region. From the point of view of knowledge production, empowering these counter-hegemonic forces is strategic."

We also observed that marginalized groups have the potential to create, in innovative ways, new spaces and knowledge-production dynamics, which challenge traditional scientific and political actors to dialogue with them and recognize their contribution to knowledge production. One relevant example is that of the Forum of Traditional Communities (*Fórum de Comunidades Tradicionais* – FCT), comprising the municipalities of Ubatuba, Paraty, and Angra dos Reis. Established in 2007, it has promoted the visibility and value given to traditional ways of living, knowledge, and cultures in that region. Its focus encompasses the relationship of communities with the earth, seasons, and crops; natural solutions for health conditions; and history, culture, and education practices of those communities. In recent years, FCT and the Oswaldo Cruz Foundation (*Fundação Oswaldo Cruz*—Fiocruz), one of the most renowned research institutions in the country, established a partnership, which created the Observatory of Healthy and

Sustainable Territories (*Observatório de Territórios Saudáveis e Sustentáveis*—OTSS). The OTSS has three axes of action: community-based tourism, sustainable sanitation, and non-standard education. Instead of being shaped by questions coming from a top-down scientific perspective, the project was based on knowledge systems, views of sustainability, and priority agendas and issues pointed out and practised by the communities themselves in their relationships with researchers and other actors.

In sum, our research indicated the existence of diverse perspectives and expectations around the roles and the methods of producing science that permeate different social groups, such as the following:

(1) responding to research demands on specific themes: the need to produce new knowledge and information relevant to social, environmental, and economic issues;

(2) improving forms of access to and appropriation of knowledge produced: how to communicate and make complex information available to different social groups, in order to positively impact the quality of citizen participation;

(3) providing information for citizen-driven monitoring, in order to empower distributed participation in data production; and

(4) developing new forms of relationships and knowledge co-production between academia, local communities, and other social groups.

Furthermore, we found that different practices of Open Science (such as Open Hardware, Open Data, citizen science, and participatory processes of engagement) are not merely isolated activities that should be lumped together under an umbrella term. On the contrary, in order to respond to local development challenges and demands, we usually need to mobilize various facets of Open Science in conjunction with one another. For example, it is not enough to provide scientific data in an open way, with the aim of responding to local issues, if this is not accompanied by ways of social appropriation of this information, which requires citizen science approaches and tools.

Additionally, there are factors that may hinder the adoption of Open Science practices in local social change. The team had difficulty in approaching and attracting well-established scientific institutions to participate in Open Science experiments. Our survey of scientific publications about Ubatuba showed that, in the last five years, about fifty percent of academic work has been published in Open Access journals. Moreover, researchers publishing in indexed journals are not necessarily those who are involved with local development issues. Universities seeking to place themselves in leading positions in international rankings give incentives to their academic and scientific staff to keep their research closed, and they disregard social issues. Career evaluation, competition over financial resources for research, intellectual property, and the need of exclusive access over primary data for "original" scientific publication are some of the key factors presented. In this regard, a professor at a university that operates a research base in Ubatuba gave the following testimony:

> There is pressure to publish, and now the number of publications has more weight than their quality. So, you direct your strategy to publish more. And then, there are some ethical problems. For example, using the same data to publish different papers, associating them with people not having an effective participation in the work. Thus the data sharing is in third, fourth priority level, because that means giving your research effort to someone else to publish without getting you involved.

Meanwhile, scientists who are more engaged with local issues seem to be more committed to Open Science values. In Ubatuba, researchers who participate in the management councils of local protected areas tend to produce a more "situated/engaged" knowledge regarding the specific issues of the territory. This made them more interested and involved in the activities proposed by the project. Moreover, these councils have demonstrated that they are not only strategic spheres of social participation due to their multi-sectoral composition and representativeness, but also relevant spaces of circulation of information and knowledge production.

We found other difficulties relating to the production and circulation of knowledge relevant to intervention in the territory, such as: (1) a lack of primary data on local problems, which is related both to little research emphasis or concern and to the lack of resources for continuous research on these topics; and (2) difficulties in organizing and making available the existing information to those participating in the debate and in public decision making on local development issues.

Notwithstanding, universities and research institutes are often called upon by managers of protected areas, community councils, and social movements to assist in the preparation of reports and analyses that can qualify their position on controversial issues. Scientific expertise is thus mobilized to provide visibility and legitimacy as a form of certification of existing social practices and community knowledge that are vulnerable to marginalization or in need of political recognition. Meanwhile, public and private companies hire those institutions to conduct scientific evaluations of the social and environmental impacts of natural resource-extraction projects. These studies, and the consequent infrastructure of dissemination created, tend to purposefully exclude the general public from accessing the knowledge generated. At the same time, public research institutions dedicated to production of applied knowledge to support government and public debate are under financial pressures (due to the contraction of public investment in science and research). It seems that a knowledge market emerges through new services that those institutions offer to the private sector, in order to obtain complementary financial resources. In some cases, it is pointed out that this practice creates new exclusivity over information access, since privileged data become a strategic asset in the competition for financial resources, and this issue has been an object of controversy among local actors.

Therefore, if the scarcity of resources may encourage the sharing of information among actors facing similar problems and issues, it can also increase the commodification of information and, therefore, its privatization. This leads to an ambiguous relationship between scientific researchers and other social groups in a context of competition for financial resources needed for institutional survival and for access to strategic information.

Moreover, there are barriers to dialogue and collaboration among different actors—particularly between non-local scientific communities and locally situated social groups—including barriers of language, of knowledge and technical skills, as well as of cultural backgrounds. The interaction between actors with different interests, worldviews, and epistemic structures implies conflicts and negotiations of distinct—often divergent—agendas, expressing asymmetric relations of power. There are significant asymmetries between those who can make use of open knowledge and collaborative practices in their interests, and those who contribute to the common knowledge but do not benefit from it. Therefore, although collaboration is a crucial part of knowledge production, it begs the question: Collaborate for what purpose, with whom, and under what terms?

We also found diverse and unequal conditions regarding access to and the use of information and communication technologies (ICTs), which are nowadays a critical requirement for information appropriation and political participation. In traditional communities, there remains a high level of exclusion from physical communication infrastructure, with many unable to rely on basic cables or antennae. Internet access is either entirely absent or expensive and unreliable, generating a serious digital divide due to low connectivity. Older community members usually have a very low level of skills with ICTs, but there are still many community leaders who handle communication applications of their smartphones quite effectively. Younger generations tend to be more technologically savvy, but generally lack money to buy data credit for their smartphones.

NGO activists and community advisory board members of Protected Areas are interested in sharing information openly, but have methodological and technological difficulties in doing so. Similarly, local government and public-sector organizations have a low level of organized information, datasets, and/or platforms for hosting open data.

Another barrier is the fragility of existing free and open software and web infrastructure with open-standard protocols for data sharing. We still very much depend on corporate tools such as Google Drive and Dropbox for the majority of online data sharing and collaboration. These tools have a high level of reliability and efficiency in the short term, and thus the adoption of free, open-source software is encumbered by more subjective obstacles such as the culture of use and inertia, informal support networks, crowd behaviour, and significant investment on user interface cross-platform usability. At the same time, high expectations around free, open-source software can be disappointing when a misconfigured service fails. Unfortunately, this is a deep-rooted and long-term threat, in a distributed context that often relies on a network of unpaid developers.

What Institutionalities?

The institutional dimension proved to be a central aspect affecting the open and collaborative nature of knowledge and its potential to bring about social change. We conceive institutions as "formal and informal rules that are understood and used by a community ... [They] are not automatically what is written in formal rules. They are rules that establish the working 'dos and don'ts' for the individuals in the situation that a scholar wishes to analyze and explain" (Hess and Ostrom 2011, 42).

Efforts in Open Science involve different levels of action and decision, ranging from individual actions to local and macro-level policies, through to the meso-level of communities, institutions, and organizations. Institutional frameworks, considered in a broad sense, include formal components such as academic recognition systems (researcher evaluation and reward criteria) and regulatory and policy frameworks, as well as informal variables such as cultural norms and larger economic aspects.

Thus, a core reflection in defining an institutional framework is the acknowledgement of how governance mechanisms (specifically those affecting knowledge and information flows) express power relations in terms of managing and resolving conflicts. This is particularly pertinent when we shift the focus toward how institutional governance mechanisms interact with development issues. Such mechanisms doubtlessly influence the means and capacity for integration, co-production, and sharing of knowledge relevant to addressing local challenges.

In Ubatuba, the local institutional environment comprises elements that both facilitate and hinder the diffusion of Open Science values and practices, and hence impact their capacity to influence positive social change. Local government and public institutions provide a legal framework that acts as a formal—and conflicted—arena in which institutions should function. On the other hand, in a less prescriptive way, the individual behaviour, attitude, and values of public and non-governmental managers—of protected areas, water resources committees, municipal secretariats, and civil society organizations—may facilitate or inhibit institutional change. In particular, as we have argued previously, within knowledge-making contexts, those managers with scientific backgrounds that are challenged and pressured to solve complex social and environmental problems are often sensitive to and interested in Open Science approaches. However, there are difficulties in committing political clout and resources to the necessary long-term support for OCS approaches and actions.

At the regional and national levels, regulatory frameworks and policies also play an important role in promoting or hindering the uptake and use of open and collaborative forms of knowledge. While such arenas are not the objects of analysis here, it is worth mentioning that an important milestone was the Brazilian *Access to Information Act* (2011) in the public sector, which has catalyzed a new field of activism and pressure for Open Access to government data, at all levels.

Access to information, combined with new and dynamic forms of knowledge production and dissemination, tends to be well aligned with the emergence of innovative institutional arrangements that interrogate dominant, top-down systems of management and traditional approaches toward undertaking scientific research. At the same time, when government institutions are embedded in a more conservative political context and are under stronger public accountability that poses a political threat to their role, they might react with information control and exclusion.

In the last couple of years, Ubatuba has opened up opportunities for citizen participation in public policy in diverse areas, which create spaces where different voices can be heard. For instance, a number of local conferences have been held in the area, and the municipality coordinated its first participatory budgeting process. Within this context, a key characteristic of the local institutional scenario is the existence of active public and multi-sectoral representation mechanisms for social participation, such as Protected Areas Management Boards formed by diverse local actors—including NGOs, traditional communities, public-sector representatives, universities, and private firms—which are in charge of suggesting and negotiating possible uses of forest, land, and marine resources. These public management mechanisms have contradictory dynamics. They actively contribute to citizen participation and local information circulation. However, their capacity to effectively intervene in the policies and decisions that affect local communities is limited, revealing the distortions of and limits to political representation. Currently, with new local and federal governments in place, it is important to follow how these multi-sectoral mechanisms will be able to act.

But even if it is possible to have an institutional framework that promotes Open Science, when it is confronted with conflicts over development, the limits on the quality (or deficit) of democratic participation become evident and constitute a barrier to the potential use of openness in building alternative forms of development. That is, even if knowledge is open and free, when considering the powers and decision-making systems in different spheres and on different scales, this openness often loses effectiveness. In other words, legislated formal equality does not necessarily imply effective equality to make use of and benefit from the results of openness in knowledge production and circulation.

Some questions remain, such as these: (1) What kinds of institutionalities do we need in order to favour social change and support open and collaborative knowledge production while protecting it from private appropriation of collective production?; and (2) How do we combine the scaling up of local social innovations in the form of knowledge production and circulation with the scaling down of macro-level policies that recognize and promote social innovation taking place on the ground? (see Smith et al. 2017.)

Concluding Remarks

A key question for our research was: Do more collaborative and open forms of scientific knowledge production open up more space for alternative forms of development? We confirmed that OCS may improve forms of co-production of knowledge, widening opportunities for vulnerable actors to influence and appropriate knowledge relevant to social and environmental demands. New spaces, objectives, and methodologies for knowledge production are conditions for alternative development and for the emergence of other modes of science production. We also verified the inverse relationship, namely the complexity of today's development challenges and the crisis of the concept and models of development that impose the need for alternative bases of knowledge and other ways of producing science.

Alternatively, we observe that the very idea of openness is under dispute (Albagli, Maciel, and Abdo 2015). We are thus left to ask "What (open) science and for whom?" Furthermore, the idea of science itself is under dispute, and this dispute lies at the core of today's democracy building. A democratic sense of openness corresponds to the enlargement of the social base and dialogue of science with other social actors. Open Science expands, or rather transcends, the so-called "scientific field" (Bourdieu 1975). In this sense, Open Science does not refer solely to the clash between public and private forms of knowledge production and appropriation, and it is not limited to increasing the speed of knowledge circulation within the field of science itself. Open Science does not concern only the potential or facility for generating or circulating information and knowledge within the so-called "scientific community" (or communities). It implies the overthrowing of hierarchies, of established sources of authority and reputation, and moving the focus to the relationship between science and power and, from a broader perspective, to the relationship between knowledge and power. Open Science is not simply a form of knowledge production and sharing that complements and democratizes decision-making on the cognitive level; it also puts into perspective the role of cognitive subjects in their respective positions of power.

Open Science further encompasses a greater permeability of science and its dialogue with other types of knowledge, considering the broad spectrum of actors, possibilities, and spaces for producing knowledge and for formulating different questions, promoting polyphony. These actors and spaces are not just "inclusive," but are mostly disruptive. Their objective is not to be included or to produce the same things in the same way mainstream science does. They create new social practices and modes of existence daily (Latour 2013) and new objectives to create knowledge, including scientific knowledge. It means that an ecology of knowledge corresponds to an ecology of powers. It involves the deconstruction of ethnocentric epistemology, valuing a lay, contextual, and situated knowledge (Haraway 1988), until now considered "subjected" knowledge (Foucault 2010).

Openness here means the struggle over a new biopolitics.[13] The possibility of a variety of ways of life constitutes the very body of a variety of ways of knowing. In other words, an ecology of knowledge corresponds to a diversity of modes of existence, an ecology of possible ways of living, of living in community (in the sense of living in common) involving other types of relations with nature, therefore, promoting different perspectives about development.

Here the dispute over the different ways of using common natural resources, such as forest, land, and marine resources, constitutes a conflictive "arena" that is intertwined with the conflicted arena over knowledge commons (see Hess and Ostrom 2011), the latter being a condition for effective decision-making and participation in them. Diverse development views are also expressed in disputes among forms of appropriation, meanings, and logistics of the use of territory and its associated knowledge. Disputes about territory are mainly reduced to its exchange value—territory as a commodity—and territory seen as a framework for living, with its multiple meanings and possibilities of use. In the same way, sustainable development is only an apparently consensual perspective. What is at stake is not merely the quantitative dimension of development—to save finite resources—but also its qualitative dimension—the use we want to make of these resources.

Thus, knowledge policies and development policies are increasingly intertwined, just as disputes and conflicts on development strategies also involve cognitive conflicts (Acselrad 2014).

Finally, one should ask "What social change?" Social change should not be understood as an equivalent to development in its hegemonic sense. From our perspective, considering the extreme inequality of our society, social change should point to the destabilization of the dominant structures of power. Therefore, Open Science should be committed to making room for counter-hegemonic knowledge, oriented to interrogate and stress the material, political, and social foundations of these inequalities and asymmetries.

From this point of view, we expanded our conceptual framework, considering the idea of common science as proposed by Lafuente and Estalella (2015), which focuses on the relationship with the diversity of modes of knowledge production. Common is used not only in the sense of common goods (the commons), but mainly common as "in between," the relationship with Otherness, with the Other. Lafuente also considers the common as the ordinary, that which is not sacred or hierarchically superior, which means that science is part of an ecosystem of modes of knowledge. It is still not possible to say to what extent the Open Science movement will contribute to the destabilization of existing scientific epistemological and institutional frameworks and practices. It implies the need for a new agenda of rights, new ethical-political issues, involving power relations between science and society.

Acknowledgment

We are grateful for the contributions received during the development of the research on which this text was based, especially from Antonio Lafuente, Henri Acselrad, Miguel Said, Lea Velho, and Cinthia Mendes, during the intermediate and final project seminars, as well as suggestions for this text received from Rebecca Hillyer and Aline Rosset.

Notes

1. We understand there are different movements under the umbrella term "Open Science."
2. This project was developed from 2015 to 2017 as part of the Open and Collaborative Science in Development Network (OCDSNet), with the financial support of Canada's IDRC and UKAid.

3. We prefer to adopt the term "local actors" instead of "stakeholders," as the latter implies a well-defined group of interests. In our research, on the contrary, we observed contradictory layers of interests within the same social group. Also, this option was reinforced by the critique in our final evaluation seminar of the use of the term stakeholder in the context of the project, since this is an approach born in the business world to refer to groups that can affect and be affected by business action, often in conflicting relationships.

4. https://pt.wikiversity.org/wiki/Pesquisa:Ci%C3%AAncia_Aberta_Ubatuba.

5. http://cienciaaberta.ubatuba.cc.

6. http://wiki.ubatuba.cc/doku.php?id=gaivotafm:radiotec.

7. https://www.youtube.com/channel/UC1J2Bd6q6VhFBNGihT2qYvA.

8. *Quilombo* is the word used to refer to the communities originating from settlements of runaway slaves in the nineteenth century. Currently it denotes communities descended from slaves that maintained aspects of their culture such as collectivism and a direct relationship to the land. In Ubatuba, there are four communities identified as *quilombolas*. However, only one of them is officially recognized as Quilombola. The others are still fighting for their land, social rights, and public/state recognition.

9. *Caiçaras* are traditional communities close to the Brazilian southeast coast. They historically made a living from fishing and farming. At the same time that they face social and cultural pressure from economic development, there is a complementary dynamic in certain regions, among new generations, to re-affirm their cultural values and practices.

10. Pre-salt oil is found underneath a thick layer of salt, in the bottom of the Atlantic Ocean. It is said to be one of the biggest oil sources in the world (see https://en.wikipedia.org/wiki/Pre-salt_layer).

11. Many of these communities have been in the region since well before the delimitation of the environmental protection areas. The Serra do Mar State Park was created in 1976. The Indigenous community of Aldeia Boa V was established in Ubatuba in 1967 and Quilombo da Fazenda was in the region before that. The caiçaras communities are still older; they have been there for over a hundred years.

12. Ecological-Economic Zoning (EEZ) is a political and technical instrument for public policies in planning the use of territory.

13. Biopolitics here is understood as the strategies of control over life as well as life itself as a form of struggle and resistance.

References

Acselrad, Henri. 2014. *"Disputas cognitivas e exercicio da capacidade critica: O caso dos conflitos ambientais no Brasil* Cognitive Disorders and Critical Capacity Exercise: The case of environmental conflicts in Brazil]." *Sociologias* 16 (35): 84–105.

Albagli, Sarita, Maria Lucia Maciel, and Alexandre Hannaud Abdo. 2015. *Open Science, Open Issues*. Brasília; Rio de Janeiro: IBICT: Unirio. http://livroaberto.ibict.br/bitstream/1/1061/1/Open%20Science%20open%20issues_Digital.pdf.

Bourdieu, Pierre. 1975. "The Specificity of the Scientific Field and the Social Conditions of the Progress of Reason." *Social Science Information* 14 (6): 19–47. https://doi.org/10.1177/053901847501400602.

Fals-Borda, Orlando. 1981. "Science and the Common People." *The Journal of Social Studies* 11: 1–21.

Foucault, Michel. 2010. "Aula de 7 de janeiro de 1976 [Class of January 7, 1976]." In *Em defesa da sociedade: Curso no Collège de France (1975–1976)* [*In defense of society: Course at the Collège de France (1975-1976)*] 3–19. São Paulo: Editora WMF Martins Fontes.

Haraway, Donna. 1988. "Situated Knowledges: The Science Question in Feminism and the Privilege of Partial Perspective." *Feminist Studies* 14 (3): 575–99.

Hess, Charlotte, and Elinor Ostrom. 2011. "A Framework for Analyzing the Knowledge Commons." In *Understanding Knowledge as a Commons: From Theory to Practice.* Cambridge, MA: MIT Press.

Herrera, Amilcar. 1972. "Social Determinants of Science Policy and Implicit Science Policy." *Journal of Development Studies* 9:19–37. https://doi.org/10.1080/00220387208421429.

Instituto Polis. 2013. "Diagnóstico Urbano Socioambiental: Município de Ubatuba [Urban Socio-environmental Diagnosis: Municipality of Ubatuba]." *Instituto Polis.* http://litoralsustentavel.org.br/diagnosticos/diagnostico-urbano-socioambiental-de-ubatuba/.

Lafuente, Antonio, and Adolfo Estalella. 2015. "Ways of Science: Public, Open, and Commons." In *Open Science, Open Issues*, edited by Sarita Albagli, Maria Lucia Maciel, and Alexandre Hannud Abdo, 27–57. Brasília: Ibict; Rio de Janeiro: Unirio. http://livroaberto.ibict.br/bitstream/1/1061/1/Open%20Science%20open%20issues_Digital.pdf.

Latour, Bruno. 2013. *An Inquiry into Modes of Existence.* Cambridge: Harvard University Press.

Morel, Regina Lúcia De Moraes. 1979. *Ciência e Estado.* São Paulo: T.A. Queiroz.

Smith, Adrian, Mariano Fressoli, Dinesh Abrol, Elisa Arond, and Adrian Ely. 2017. *Grassroots Innovation Movements.* London; New York, NY: Routledge.

Sunkel, Osvaldo, and Pedro Paz. 1970. *El Subdesarrollo Latinoamericano y la Teoría del Desarrollo* [Latin American Underdevelopment and Theory of Development]. México: Siglo XXI.

Varsavsky, Oscar. 1972. *Hacia una política científica nacional* [Towards a National Scientific Policy]. Buenos Aires: Ediciones Periferia.

CHAPTER 14

Toward African and Haitian Universities in Service to Sustainable Local Development: The Contribution of Fair Open Science

Florence Piron, Thomas Hervé Mboa Nkoudou,
Marie Sophie Dibounje Madiba, Judicaël Alladatin,
Hamissou Rhissa Achaffert, and Anderson Pierre

Abstract

This chapter is a contribution based on our action-research project on Open Science in Francophone Africa and Haiti (called SOHA project). The project was led by a large group of scientists, researchers, and students of all levels, representing about fifteen countries. For over two years, this group has been thinking about obstacles to the adoption of Open Science in Francophone Africa and Haiti and about the invisibility of researchers from this area of the world in the world's scientific conversations. Initial results of this study are presented in two parts: the first part gives an account of our work on neocolonialism and cognitive injustices that are rife in African and Haitian universities. In the second part, we present avenues of appropriation of Open Science in African and Haitian contexts, and we propose concrete solutions so that their universities may be of service for local, sustainable development.

Introduction

Being inspired by scholars such as Keim (2010) and others (Kreimer 1998; Polanco 1990; Vessuri 1994) and by Wallerstein's theory (1996), we consider that science has been historically globalized and constitutes a world system organized around scientific publications. Produced mainly in the North, this merchandise obeys standards and practices that are defined by the "centre" of the system, namely the main commercial scientific publishers (Larivière, Haustein, and Mongeon 2015) and their partners from US and British universities dominating the so-called world rankings. The semi-periphery is constituted by all the other countries of the North or emerging from the South, which revolve around this centre with English as the primary language of science. The periphery then refers to all the countries that are excluded from this system, which produce very few scientific publications or whose research is invisible (Charlier, Croché, and Karim 2009; Hountondji 2001). With a contribution of less than one percent of the world's scientific publications, Sub-Saharan Africa belongs to the periphery (Nwagwu 2013; Piotrowski 2014; Kotecha, Wilson-Strydom, and Fongwa 2012; Mboa Nkoudou 2016).

While many international reports consider higher education and scientific research as development tools (Crossley and Watson 2003), many questions arise: considering this "apparent inexistent scientific production," how can Sub-Saharan universities contribute to the development of their countries? Which science are we talking about? Which development is it? Which strategies do African universities need to adopt to ensure the development of their countries?

In this chapter, we tackle these issues by presenting some findings of the SOHA research-action project. For this purpose, we have three core objectives: (1) identify the invisibility of African scientific publications by describing cognitive injustices, (2) to make a theoretical clarification on Open Science and development, and (3) propose concrete solutions for the adoption of Open Science by African universities. Before presenting these findings, we will describe the SOHA project and its methodology, which led us to these conclusions.

The SOHA Project Methodology

"Project SOHA" was a research-action project working on Open Science, empowerment, and cognitive justice in French-speaking Africa and Haiti from 2015 to 2017. The choice of these areas of the world can be explained by the fact that, even inside the periphery, Francophone Sub-Saharan African universities seem non-existent with only 0.01% contribution in the world's scientific publications (Mboa Nkoudou 2016), and reports on the state of scientific research in Haiti after the 2010 earthquake (Machlis, Colòn, and McKendry 2011) indicate the same situation. On the other hand, the connection between Haiti and Francophone Sub-Saharan Africa is also cultural and historical.[1]

The work of the SOHA project was directed toward understanding the views of students and researchers enrolled in various public universities in Francophone sub-Saharan Africa and Haiti. At the early stage of the project, a Facebook group was set up to recruit new members to the project, to connect researchers, to share information, to facilitate collaboration and communication, and to discuss issues of interest, etc. This group is a great source of qualitative data for our research; today it is still active with more than ten thousand members from Africa and Haiti. With SOHA members represented on all campuses in Haiti and Francophone Africa as a result of our Facebook group, a questionnaire was administrated physically and online (Google forms) for those who could afford Internet service. Besides the questionnaire and interactions through the Facebook group, data were also gathered by inviting students to write blog posts; conducting numerous group chats on Facebook, Messenger, and WhatsApp; taking part in collaborative writing; and organizing seminars and symposia in Haiti, Burkina Faso, and Cameroon. All these sources of information allowed us to collect qualitative and quantitative data, which has been analyzed; some results have already been presented in scientific papers and also published in a book entitled *Justice cognitive, libre accès et savoirs locaux: Pour une science ouverte juste, au service du développement local durable* [Cognitive Justice, Free Access and Local Knowledge: For fair open science at the service of sustainable local development] (Piron 2016). In this chapter, we are presenting additional conclusions, which will allow us to deeply understand how openness should be contextualized in higher education to contribute to development.

Findings: Cognitive Injustices

We have given a new meaning to a concept originally intended to qualify the aspiration for active recognition of the plurality of knowledge in science: *cognitive justice* (Visvanathan 2009). We now define cognitive justice as an epistemological, ethical, and political ideal aimed at the creation of socially relevant knowledge across the globe (not just in the North), within a science-practising inclusive universalism, open to all forms of knowledge. From this perspective, we consider the difficulties faced by African and Haitian scholars and students to carry out and publish research, as cognitive injustices reducing their ability to deploy the full potential of their intellectual skills, their knowledge, and their scientific research capacity to serve sustainable local development of their city, their region, and their country. This concept of cognitive injustice allows us to achieve the objective of this chapter: identify the invisibility of African scientific publications. To illustrate this, we present a list of nine cognitive injustices experienced by students and researchers from French-speaking African countries and Haiti below.

Cognitive Injustice 1: Infrastructure and Research Policies Are Lacking in Africa and Haiti.

African and Haitian (public) universities very rarely have the financial, administrative, and informational resources required to develop a viable system of scientific research, which includes laboratories, equipped libraries, universal internet access, research centres, funding agencies, scientific journals, etc. Rather, our investigation shows administrative difficulties for young scientists, the lack of science policy across the country, minimal salaries for teachers, and a dependence on Northern countries for research grants. Disciplinary divisions and rivalries between faculties and between senior administrators do not help to create a favourable working environment for research. How can knowledge be produced in these conditions, if not at the cost of personal sacrifice? Only a true political will in every country can reverse this situation.

Cognitive Injustice 2: Access to Scientific Publications Is Often Closed.

While they are the main source of references in scientific research, most scientific articles on the web are not accessible to potential readers. This phenomenon goes unnoticed in the eyes of those who are affiliated with a university whose library can afford to subscribe to scientific journals that publish these articles, notably in Northern countries. On the other hand, people who are not affiliated with a university or those whose university is too poor to subscribe to these journals only have access providing they pay a certain amount typically with a credit card. In Haiti or Africa, very few people possess a credit card, especially among students. These people are deprived of access to scientific resources that are necessary to produce high-quality research.

The Open Access movement proposes an answer to this injustice. It encourages scientists from Northern countries and the Global South to publish in Open Access journals (i.e., journals that make research freely available to the public) or archive a digital copy of their texts in an institutional digital repository, making them available online for all to access. Despite the resistance of commercial scientific publishers (Elsevier, Springer, De Boeck, etc.) and a certain conservatism among scientists, the movement toward Open Access seems irreversible, as evidenced by the recent policy of Canadian granting agencies and the European Union (Piron and Lasou 2014).

In terms of web archiving of dissertations and theses—a huge pool of valuable knowledge that is rarely published—great progress has been made in some Northern countries, which generally integrate them into their Open Access policy. For example, at Laval University, students must submit a digital version of their thesis and accept that it be available in Open Access as a condition of graduation. Awareness is growing in the Global South, but making MA theses, PhD theses, and research work, which tends to languish on shelves rather than being available to everyone on the web, universally accessible could go much faster. In 2016, Senegal repatriated four hundred theses by Senegalese researchers (Sylla 2016).

Cognitive Injustice 3: Digital Literacy and Access to the Web Are Rare.

Our research project clearly confirmed both the difficulty of access to the web for university students and academics of Francophone Africa and Haiti, as well as their low rate of digital literacy. Digital literacy refers to the ability to optimally exploit the potential of a computer and the web. For example, some students only touched a computer for the first time during their first year of university. Many have no email address or use the computer merely as a typewriter. They often have no idea of free scientific and educational resources that are already available on the web, while Northern universities introduce these to their students and teachers.

Causes of limited access to the web and low digital literacy are not simply due to a lack of financial resources for universities or countries. This is largely the result of political choices, web access not being a priority for many universities in Africa and Haiti. Yet providing universal access to the web on the campuses of African and Haitian universities would be a very effective action to counter the low digital literacy rate and enable university students and academics to learn the essential digital skills required to disseminate scientific and technical information globally. With such access, university students and academics could best use the resources of the free scientific web and improve the quality of teaching and research. They may also be less tempted by the "brain drain" if they found the same access to international scientific dialogue as in their country.

Cognitive Injustice 4: Local Knowledge Is Excluded or Disrespected.

In the positivist-normative framework that dominates current science, knowledge that is local, oral, practical, experiential, or contextual is considered not to be knowledge to be either ignored or retranslated in scientific terms by experts. Students have expressed great anxiety because they felt they have had to give up on their local knowledge to embark on science—a knowledge cherished and valued by their family and friends. Even a small mention of the inherent value of all knowledge, including local knowledge, helped students to identify, in a positive way, with the SOHA project (Achaffert 2015; Mboa Nkoudou 2015; Pierre 2016).

Cognitive Injustice 5: The Wall Between Science and Society Is a Barrier.

On behalf of the positivist ideal that science implies neutrality but also fearing external interference in science that would make it "impure," and hence less scientific, scientists in all countries are trained to distrust all that is political and refuse to make scientific and research processes accessible to non-scientists, whether they be in industry, political power, or civil society. Unfortunately, this position generates an isolated science cut off from society and deprived of the support of citizens who do not understand its purpose. This position also harms the eventual political will to make science and university tools for sustainable local development. Yet many scientists from Haiti and Francophone Africa are motivated mainly to improve their country's situation.

Cognitive Injustice 6: The Western Research-Publication Model Is Closed.

The Western publishing system, dominated by Anglo-Saxon journals, is very demanding for researchers from the South wishing to publish. Far from encouraging the diversity and quality of knowledge, it is based on competition between researchers who must "publish or perish" and on the fight for scarce publishing venues. Peer reviewing aims to eliminate articles that do not match the criteria of excellence set by journals with increasing homogeneity. It is even more difficult for French-speaking researchers from the Global South to succeed in this system, yet publication in these elite journals is a criterion for promotion in universities of Francophone Africa. The solution to this problem is twofold: on one hand, the vigorous debates that animate Anglo-Saxon science may soon lead to a questioning of this system. On the other hand, some SOHA members are currently working on the *grenier des savoirs* project (attic of knowledge), an Open Access scientific publication system for and by researchers from Africa and Haiti.

Cognitive Injustice 7: The Language of Science Is Colonial.

The dominance of Anglo-Saxon commercial publishers and their control over the databases from which the journal impact factor is calculated, reinforces the hegemony of the English language on science, while claiming universality. For scientists from Francophone Africa and Haiti, places whose colonial language was French, English poses a significant barrier, particularly in the form of written, academic text. Since scholars in post-colonial situations tend to speak at least one national language (their mother tongue) as well as French, English becomes their third or fourth language. How can a person work, think, and produce knowledge to the best of their abilities when one must use a language that they have not yet mastered?

For language equity, scientific publications could open themselves to multilingualism. Without giving up English or French as a contact language, journals could encourage authors to write in the language of their choice. They should jointly publish the original text and translations in different languages. This is what we wish to accomplish with our publishing house *Éditions science et bien commun* (http://editionscienceetbiencommun.org).

Cognitive Injustice 8: The Pedagogy of Humiliation Is Still Rife.

We have received numerous testimonies of an unfortunate pedagogy in universities, especially at the master's and doctorate levels: professors who transform their positions of power into the right to destroy those who might replace or surpass them by practising the "pedagogy of humiliation." Public humiliation in the classroom, refusal to read student works, stringent assessments, thesis defence postponed for months, destructive criticism — suffering this can only block the potential of future scientists in these countries. Recognizing this type of pedagogy, encouraging doctoral students to refuse or resist it, and, above all, showing that another pedagogy focused on empowerment is possible — this is what can be done. This cognitive injustice may be reinforced by the obligation of respect for authority, hierarchy, and elders, often present in traditional African societies.

Cognitive Injustice 9: Epistemic Alienation Is Profound.

Postcolonial research, including the works of Fanon (2002), has shown that the colonization of minds has accompanied that of the body and the territory. Quijano (2000), Thiong'o (2011), and others propose to decolonize the thought and knowledge of the Global South by criticizing the universalist pretensions of modernity and showing its very localized presence in Europe. Scientifically, the project of this "decoloniality" is the deconstruction of positivism and of its hegemony on contemporary science, as well as the enhancement of epistemologies or ways of knowing specific to the Global South.

These cognitive injustices are reinforced by the financial dependence of Sub-Saharan Francophone researchers on funders from the North who can either hire them as local researchers for their projects or support local projects that correspond to their priorities for action. Those "partners" inevitably orient the constitution of the problems and the methodological and epistemological choices of African researchers toward the only model they know and value, the one born at the centre of the global system of science, without questioning whether this model is relevant to Africa and its challenges. Hountondji describes this extroversion very well: despite government declarations of intent (Irikefe et al. 2011; Kigotho 2014; Nordling 2010), "postcolonial scientific research remains fundamentally extroverted: outward-looking organized to meet a demand (theoretical, scientific, economic, etc.) that comes from the World Market Center" (Hountondji 2001, 4). This lack of financial resources means that African universities cannot afford to meet the development needs of their country. Starting with the origin of this devastating assessment, Fredua-Kwarteng (2015) explained the gap between African universities and development:

> Over the decades, African universities, particularly the publicly funded ones, have played a significant role in developing human resources for state bureaucracies including ministries, departments, boards and agencies, the education sector and the professional class, such as lawyers, bankers, judges, engineers, doctors, accountants and managers. Nonetheless, African universities have had minimal to zero impact on producing the people who can solve the developmental problems plaguing the African continent. In fact, graduates emerging from universities tend to perpetuate the status quo rather than transform the state organizations that employ them. They are imbued with a colonial sense of entitlement, lack problem-solving skills and demonstrate low levels of work productivity.

In other words, post-colonial African universities are considered by African governments as machines for producing and reproducing the countries' elite and their social order, and not as places where new ideas and new knowledge can be created to help solve the most pressing problems of the people. Therefore, despite the respectable number of researchers, the research work carried out only partially satisfies the need, and is hardly commensurate with the great anxiety felt by victims of Francophone African states of an increasing marginalization within the global economic system.

All these cognitive injustices mean that scientists from Francophone Africa and Haiti must think and research without having the material and financial means, in a language that is not their own, and in an epistemology that they inherited from colonization and that leads them to devalue local knowledge and ways of knowing.

That is why it is so important to refine ideas like Open Science and development.

"Fair Open Science" as a Tool of Local Sustainable Development

From the beginning of the SOHA project, we experienced openness in choosing to address the links between science and sustainable local development in Francophone Africa and Haiti not from a "neutral point of view," but from a viewpoint located in real contexts. We believe—and are trying to confirm—that our concept of Open Science, which goes beyond Open Access to scientific publications or even the participation of non-scientists in research projects, may instead help scientists from the Global South to deploy their potential for creating knowledge for sustainable local development, including students who are not yet completely under positivist tutelage or who challenge it. We call this brand of Open Science "fair Open Science" and believe it can be used to explore burning questions such as: How can they better contribute to sustainable local development in their country? How should students become empowered to create locally relevant knowledge, despite infinite difficulties in their daily lives? But first, we should be careful about the neo-colonial face of openness.

The Neocolonial Face of Open Access

The movement of Open Access to scientific publications, born in universities in the Global North in the 1990s, is not devoid of ambiguities regarding its aims. It is possible to identify several aims within the various arguments used by its leaders. First, Open Access can have the objective of increasing scientific productivity and quality. For example, Eysenbach (2006), finding that Open Access maximizes the number of citations of an article, concludes that "OA is likely to benefit science by accelerating dissemination and uptake of research findings." Indeed, Open Access to publications and scientific data facilitates and accelerates the flow of research results and protocols, which can avoid duplication and unnecessary replication, etc. Needless to say, this purpose is perfectly in place within the normative framework of the dominant positivist science.

Another possible objective of Open Access is primarily economic: "Open Access to science and data equals cash and economic bonanza," said Neelie Kroes, Vice-President of the European Commission (Kroes 2013). Why? Because it supposedly facilitates innovation by increasing the flow of information and the sharing of risks. In this case, Open Science is a strategy at the heart of the knowledge economy in its open innovation variant (De Backer 2008). The networks of living labs and fab labs manage to combine participatory open innovation and incubators of for-profit business. Finally, one can identify a third official purpose of Open Access: the democratization of access to science to different audiences who do not have easy access to university library resources, including pre-university teachers, non-scientists, and organizations of civil society. In turn, it allows them to contribute to scientific knowledge, for example through participatory science (Bustamante 2015).

Contextualizing these three purposes within universities in Francophone Africa and Haiti leads us to rethink them deeply. Indeed, in these countries, funding for scientific research, good stable salaries for academics, and good quality web access are rare, while digital illiteracy is common in academia. In this context, designing Open Access as a way to maximize the efficiency and productivity of the scientific research process is meaningless. The scientific research process must first be truly launched in many of those countries where there are neither scientific journals nor research centres or grant programs to support scientists. Moreover, the country rankings in scientific production published by scientific platforms such as Scopus and Web of Science show that French-speaking Africa produces less than 0.01% of world scientific production (Mboa Nkoudou 2016). Although one can contest the validity of these rankings that ignore local scientific works and those in French, the fact remains that the world science is essentially in the North and that competition issues between laboratories and scientific productivity primarily concern these countries (Piron et al. 2017).

Similarly, the economic purpose of Open Access defended by advocates of the knowledge economy who are constantly in search of marketable innovations seems irrelevant in a context where the formal economy and industrialization are stagnant. There are other priorities in the Global South than the fight for Open Access to Elsevier journals, for example. Seen from the Global South, this fight implicitly involves easy access by well-paid researchers to basic digital tools, research infrastructures, and research grants that can even pay the exorbitant fees charged by some journals in the name of Open Access (Hachani 2014).

However, the purpose of democratizing access to knowledge seems crucial not only for non-scientists, as in the North, but for professors and students, who in Africa and Haiti are in a chronic condition of lacking access to up-to-date, high-quality scientific and technical information. Indeed, as confirmed by our survey (Piron and Mboa *forthcoming*), Francophone African and Haitian academic libraries lack financial and documentary resources, a situation that undermines their mission to improve the conditions of study and work of students. Every time a scientist from the North makes their work openly accessible, he or she makes it available not only to peers, civil servants, teachers, businesses, and organizations in his or her country, but also to all students from Africa and Haiti, to the extent that access to the web allows them to download these papers. However, scientific studies published in journals from the North are mainly authored by scientific authors from the North, thinking in a Northern epistemology and probably working on research questions that reflect local issues from the North and scientific policies of these countries. If Open Access is limited to facilitating the access of scientists from the Global South to this science from the North, it will do nothing but increase their epistemic alienation, their habit of referring primarily to science from the North. This can strengthen the difficulties of creating a locally relevant and meaningful science, using epistemic frames adapted to the context of use and in a language they can understand and use. These difficulties are discussed here through the concept of cognitive injustice.

Which Epistemology for Which Development?

The concept of "development" has long been subject to much criticism, especially because of the Western-centric and imperialist dimension of this vision of the Global South (Latouche 2001). The current hegemony of neo-liberalism and managerial thinking in the North encourages an obsession with economic development among major international organizations and their experts. Yet its obvious failure is continually proven by the appalling global inequalities that persist between countries of the North and the Global South, especially Francophone Africa and Haiti.

The idea of "post-development" theorized by Escobar (2000, 2007; Ziai 2007) offers a different interpretation of the divide between North and South. Instead of seeing it as a sign of backwardness of the countries in the Global South compared to a supposedly universal standard embodied by the North, the idea signals the difficulty of some countries to develop according to their own priorities, norms, and values, in their language and with respect for their environment. We call this other type of development "local or community development." From our point of view, its huge advantage is to take for granted the necessary empowerment of people in their territory or living environment. From that perspective, we interpret the divide between the Global North and the South as an injustice between some nations (in the North) who were able to develop according to their values and priorities (hence grow locally) and countries that failed to do so (in the Global South). One of the reasons for this failure is that the standards of most powerful countries remain dominant and colonize the futures of others by exploiting their natural human and material resources.

In the current context of global warming, which affects the entire planet, local development cannot be isolated from the rest of the world. It must be part of the global struggle for the preservation of the environment and of the natural resources necessary for life on Earth. It must also be part of the search for an alternative option to the neo-liberal model of economic growth that harms the environment. We add the "sustainable" adjective to describe this vision of local development to which we adhere.

The critique of imperialism inherent in the prevailing concept of development based on economic growth does not only show that it is a tool of exploitation and oppression of the Global South, but it also targets the conviction that this development model must apply everywhere in the same way, it being the only possible and thinkable model. In other words, the "singular" development in the dominant theory erases all possibilities of plural. That is to say that it eliminates a plurality of forms and types of development. Yet this diversity is essential to the idea of sustainable local development: development models vary according to local contexts and issues.

The unitary discourse of development is obviously the fruit of modernity (Sarr 2016). Modernity is also defined by its effort to bring about an epistemology centred on the quest for "the" truth embodied in the scientific project (Foucault 2001). The singular(ity) of that truth is delimited by a definite article ("the") and establishes what Foucault calls a regime of truth (Foucault and Gordon 1980) built on the exclusion or lesser value placed on other truths or other knowledges. For instance, traditionally excluded knowledge may include theoretical and empirical knowledge produced in peripheral universities that criticize the western-centrism of science, Indigenous knowledge, sacred knowledge transmitted secretly during rituals, knowledge specific to practices, gender, or age (women's knowledge, elders' knowledge, men's knowledge, etc.), but also political knowledge (knowledge of oppression, social memory, collective memory), experiential (subjective) knowledge, and knowledge astride the border between art and science, and culture and science. We call all these knowledges "local knowledge" to indicate that they are related to human experiences localized in contexts and do not have the aim of generalization or contextualization that marks the scientific kind of knowledge. Our epistemological position is that local knowledge is knowledge because it allows a multitude of social actors to interpret the world and act in it. In the scientific field, the singular continues to dominate, giving birth to the idea of "scienticity," that is to say, a set of material requirements and cognitive criteria allowing the assessment of whether a particular knowledge may or may not be considered scientific and officially enter the pool of knowledge that constitutes "science." Among those criteria are the generalized dimension of produced knowledge, the publication of this knowledge (called "research" in English) in a core academic journal after peer review, the use of standard research methods, a doctorate degree and an academic position for its author, the choice of an English-speaking journal with high-impact factor, etc. The "evidence-based" criteria belong in this semantic universe.

We call this unitary epistemological approach "positivism." We view this approach not only as exclusionary to the plurality of human knowledges from the field of science, but see it as imposing only one specific way of doing science: alignment with the positivist legal framework and its definition of what is scientific. For fifty years, the constructivist critical, anarchist, feminist, and post-colonial science (Berger and Luckmann 1967; Feyerabend 2010; Harding 2004; Harding 2011; Thiong'o 2011) has showed that this generalization of science was itself a local knowledge, rooted in a history, institutions, interests, and values associated with modernity and colonization, although claiming to be the only truth. This knowledge rooted in the history of the West has features that make it especially powerful: its power to tell the truth and so to impose a truth (Foucault 2001), its ability to think about how it is applied, disseminated, and challenged, and its ability to synthesize and integrate a huge number of diverse knowledges in the movement of knowledge creation (systematic review). Our critical stance does not lead us to reject science per se, but rather to reject its claim to be the only way to recognize alternative knowledges. We call for a radical transformation of this claim.

In sum, the critique of "development" leads to a necessary epistemological critique: Western science and scientists are often unable to think in the plural, to open up to a plurality of knowledges and epistemologies, especially from the Global South. Except in certain practices of social sciences, the dominance of the positivist epistemology prevailing in the global network of universities immediately disqualifies local research topics or topics expressing an interest in local issues because they would be too "engaged" or not quite generalized enough. This cursory argument has shown the links between positivist epistemology, the knowledge economy, the changing role of universities in the North, and the perpetuation of the development model based on economic growth. This hegemonic model, defined as the legacy of modernity, claims to be the only possible model of science for development in the North as in the South. The *University World News* is the perfect vehicle for these ideas.

Promoting the empowerment of scientists in developing their power to act and build locally relevant and useful knowledge is to allow Open Science to contribute to the development. In the next section, we explore strategies for the contextualization of Open Science.

Strategies to Contextualize Open Science in Francophone and Haitian Universities

To contextualize Open Science in Africa, we propose to add the obligation to consider the local situation of the scientists working in the Global South and their situation of cognitive injustice, instead of imagining that science is universal and works in the same way everywhere. This contextualization of Open Science should be accompanied by a strategy of empowerment of scientists from the Global South in order for them to meet the challenges of their local development. From this point of view, we can engage African and Haitian universities in two ways: make the research conducted in African and Haitian universities visible and connect these searches to the needs of local people.

In terms of visibility, Open Access is an interesting option for researchers from the periphery to participate in the great scientific conversation. However, this conversation should not only be from the North to the South but should focus on the visibility and increase of knowledge from the South, so that the world becomes aware of its huge wealth. If we advocate Open Access publication, we consider that the practice of APCs (Author Publication Charges) by for-profit publishers is intolerable, as are the science policies from the North that allow these APCs to accommodate this industry. An epistemology of solidarity demands three actions concerning Open Access: that the journals are supported by public funds and managed by academics; that universities increasingly support open archives; and that scientific publication is freed from lucrative issues thanks to free software, for example (Piron et al. 2017). Similarly, we must encourage the adoption of open archiving in Francophone African universities.

However, we do not advocate for the proliferation of institutional repositories in Africa and Haiti because they are too expensive to maintain. Considering that each university should have its own institutional repository is part of the logic of the North where universities compete for visibility and world ranking. Our proposal is to create a pan-African institutional repository, which could accommodate the scientific production of several universities in Africa and Haiti. In January 2018, this project was officially endorsed by CAMES (*Conseil africain et malgache pour l'enseignement supérieur* [African and Malagasy Council for Higher Education]). This is fair Open Science in action and not just Open Science.

In order to better connect African research to the needs of local populations, we discussed the creation of science shops as a means to build links between a university and local civil society toward sustainable local development. This is only one way by which we can extricate ourselves from the positivist stance that ignores the plurality of knowledges and contexts and advocates an indifference to the contexts and local issues, considered as a threat to the generalization endeavour. On the contrary, our concept of Open Science invites scientists to come out of their "confined laboratory" toward "outdoor research" (Callon, Lascoumes, and Barthe 2001), working with non-academic social actors on action research projects, joint research, applied research, or industrial research. In other words, rejecting the ivory tower, practitioners of Open Science agree to be involved in the life of their community. In college, fair Open Science not only demands more collaboration between faculty, students, and civil society or the job market, but also promotes opportunities like science shops that link civil society organizations with students and professors to conduct research or to do practical projects jointly (Piron 2009, 2016; Leydesdorff and Ward 2005). A university that sets up a science shop strengthens the mission and capacity of local civil society organizations and sensitizes students to citizen engagement. The practice of Open Science thus leads a university to develop a concept of development more oriented toward local issues and active participation of local civil society to set collective priorities. Note that this option also requires a pedagogical transformation toward practical projects that can benefit the development of students' abilities and their involvement in local issues. To further these ideas, Piron now leads a five-year project on science shops in Haiti, Africa, and India.

Conclusion

Could the adoption of fair Open Science in African and Haitian universities lead to increased possibilities and tools for sustainable local development? Without minimizing the difficulties associated with such an objective, which the SOHA project aimed to document, we have sought to identify the aspects of fair Open Science that could influence these universities on several crucial points.

We believe that a university that chooses to highlight the knowledge produced by its students and faculty members, for example by creating an open digital archive of theses and articles locally produced, is a university that will actively contribute to sustainable local development. For such an archive to be useful and used, it must be accompanied by a free Wi-Fi network on campus, as well as various financial and technical resources to support science enthusiasts. But it should also include the recognition of the value of local knowledge and local languages. This recognition can have many effects in the fight against epistemic alienation and against the imposition of colonial languages as the most legitimate and scholarly languages.

Our collaborative work made us realize that in order to become a sustainable local development tool and not a tool of neo-liberal development, Open Science must be "fair." This means it must take into account the context of cognitive injustice in which students and researchers from the Global South must work—a context that prevents and dissuades them from generating relevant, local knowledge. A seemingly "neutral" Open Science, impermeable to the contexts in which scientists work, is automatically on the side of the strongest, that is to say, the neo-liberal, positivist dominant system of science. In contrast, fair Open Science aims to develop the capacity for scientists of the Global South to think, to search, and to publish valuable local knowledge. A university that makes this choice protects its collective capacity from risks of enclosure and diversion to the decontextualized knowledge economy. It makes its commitment to sustainable local development because the ability to innovate in the service of the common good is enhanced. Therefore, despite the obstacles identified by the SOHA project, it seems to us that the practice of fair Open Science in African and Haitian universities could help their transformation into tools that are in harmony with local development priorities.

Notes

1. Indeed, the Haitian population is mainly composed of descendants of slaves, many of whom came from Benin, with whom Haiti also shares a common religion, *voodoo*. Beyond this cultural aspect, Haiti and Francophone Sub-Saharan Africa share a similar colonial history, as they were all colonized by France.

References

Achaffert, Hamissou Rhissa. 2015. *"Le paradoxe d'une science dite 'universelle'* [The paradox of a so-called universal science]" *Blog du Projet SOHA*. November 23, 2015. http://www.projetsoha.org/?p=885.

Berger, Peter L., and Thomas Luckmann. 1967. *The Social Construction of Reality: A Treatise in the Sociology of Knowledge*. New York, NY: Anchor.

Bustamante, Maria Constanza Aguilar. 2015. "Open Access and Changes in Democratization of Access to Knowledge Policies." *Diversitas: Perspectivas en Psicología* 11 (1): 9–10.

Callon, Michel, Pierre Lascoumes, and Yannick Barthe. 2001. *Agir dans un monde incertain — Essai sur la démocratie technique* [Acting in an uncertain world - Essay on technical democracy]. Paris : Seuil.

CAMES. 2013. "Premières journées scientifiques du CAMES: Quelles stratégies de fonctionnement, développement et financement durables de la recherche scientifique en réseau dans l'espace CAMES?" CAMES. http://www.agrhymet.ne/portailCC/images/pdf/cames2.pdf.

Charlier, Jean-Emile, Sarah Croché, and Ndoye Abdou Karim. 2009. *Les universités africaines francophones face au LMD : Les effets du processus de Bologne sur l'enseignement supérieur au-delà des frontières de l'Europe* [Francophone African universities face the LMD: The effects of the Bologna process on higher education beyond the borders of Europe]. Louvain-La-Neuve : Editions Academia.

Crossley, Michael, and Keith Watson. 2003. *Comparative and International Research in Education*. London: Routledge, https://doi.org/10.4324/9780203452745.

De Backer, Koen. 2008. *Open Innovation in Global Networks*. Paris: Organisation for Economic Co-operation and Development.

Escobar, Arturo. 2000. "Beyond the Search for a Paradigm? Post-Development and Beyond." *Development* 43 (4): 11–14. https://doi.org/10.1057/palgrave.development.1110188.

———. 2007. "Post-Development as Concept and Social Practice." In *Exploring Post-Development: Theory and Practice, Problems and Perspectives*, edited by Aram Ziai. London: Routledge.

Eysenbach, Gunther. 2006. "Citation Advantage of Open Access Articles." *PLoS Biology* 4 (5): e157. https://doi.org/10.1371/journal.pbio.0040157.

Fanon, Frantz. 2002. *Les damnés de la terre*. Paris: Découverte/Poche.

Feyerabend, Paul. 2010. *Against Method*. 4th ed. London: Verso.

Foucault, Michel. 2001. *Dits et écrits, tome 4*. Paris: Gallimard.

Foucault, Michel, and Colin Gordon. 1980. *Power-Knowledge: Selected Interviews and Other Writings, 1972–1977*. New York, NY: Pantheon Books.

Fredua-Kwarteng, Eric. 2015. "The Case for Developmental Universities." *University World News*. http://www.universityworldnews.com/article.php?story=20151028020047530.

Hachani, Samir. 2014. "Politique(s) du libre acces en Algérie: état des lieux et perspectives," paper delivered to the international conference on Open access and scientific research: towards new values, Tunis, Tunisia http://icoa2014.sciencesconf.org/36289.

Harding, Sandra. 2004. *The Feminist Standpoint Theory Reader: Intellectual and Political Controversies*. Hove: Psychology Press.

———. 2011. *The Postcolonial Science and Technology Studies Reader*. Durham, NC: Duke University Press.

Hountondji, Paulin J. 2001. "Le savoir mondialisé : déséquilibres et enjeux actuels [Globalized Knowledge: Imbalances and Current Challenges]." *La mondialisation vue d'Afrique, Université de Nantes/Maison des Sciences de l'Homme Guépin*.

Irikefe, Vivienne, Gayathri Vaidyanathan, Linda Nordling, Aimable Twahirwa, Esther Nakkazi, and Richard Monastersky. 2011. "Science in Africa: The View from the Front Line." *Nature*, 474 (7353): 556-59. https://doi.org/10.1038/474556a.

Keim, W. 2010. "Pour un modèle centre-périphérie dans les sciences sociales." *Revue d'anthropologie des connaissances* 4 (3): 570-98.

Kigotho, Wachira. 2014. "Dilapidated Universities Must Be Fixed, Say Officials." *University World News*, 343. www.universityworldnews.com /article.php?story=20141113165240708.

Kotecha, Piyushi, Merridy Wilson-Strydom, and Samuel N. Fongwa. 2012. *A Profile of Higher Education in Southern Africa–Volume 2: National Perspectives*. SARUA.

Kreimer, Pablo. 1998. "Understanding Scientific Research on the Periphery: Towards a New Sociological Approach?" *East Review* 17 (4): 13–22.

Kroes, Neelie. 2013. "Open Access to Science and Data = Cash and Economic Bonanza." *Europa.eu*. Novembre 19, 2013. https://ec.europa.eu/digital -single-market/en/news/open-access-science-and-data-cash-and -economic-bonanza.

Larivière, Vincent, Stefanie Haustein, and Philippe Mongeon. 2015. "L'oligopole des grands éditeurs savants [Oligopoly of the Great Scholarly Publishers]." *Découvrir: Le magazine de l'Acfas*. http://www.acfas.ca /publications/decouvrir/2015/02/l-oligopole-grands-editeurs-savants.

Latouche, Serge. 2001. "En finir, une fois pour toutes, avec le développement." *Le Monde diplomatique*. https://www.monde-diplomatique.fr/2001/05 /LATOUCHE/1754.

Leydesdorff, Loet, and Janelle Ward. 2005. "Science Shops: A Kaleidoscope of Science–Society Collaborations in Europe." *Public Understanding of Science* 14 (4): 353–72. https://doi.org/10.1177/0963662505056612.

Machlis, Gary, Jorge Colón, and Jean McKendry. 2011. *Science for Haiti, A Report on Advancing Haitian Science and Science Education Capacity*. AAAS.

Mboa Nkoudou, Thomas Hervé. 2015. "Le projet SOHA nous a tendu la main. . . et nous l'avons saisie." *Blog du Projet SOHA*. May 8, 2015. http:// www.projetsoha.org/?p=36.

———. 2016. "Le Web et la production scientifique africaine: visibilité réelle ou inhibée?" *Site du projet SOHA*. June 8, 2016. http://www.projetsoha .org/?p=1357.

Nwagwu, Williams E. 2013. "Open Access Initiatives in Africa—Structure, Incentives and Disincentives." *The Journal of Academic Librarianship* 39 (1): 3–10. https://doi.org/10.1016/j.acalib.2012.11.024.

Pierre, Anderson. 2016. "La science ouverte, une bouffée d'air pour les savoirs traditionnels en Haïti? Lasyans altènativ, yon mwayen pou valorize konesans tradisyonèl ann Ayiti?" *Blog du Projet SOHA*. January 17, 2016. http://www.projetsoha.org/?p=947.

Piotrowski, Jan. 2014. "Top Development Journals Dominated by Northern Scholars." *SciDev.Net*. www.scidev.net/global/publishing/news /development-journals-northern-scholars.html.

Piron, Florence. 2009. "Les boutiques de sciences." *Aux sciences, citoyens! Expériences et méthodes de consultation sur les enjeux scientifiques de notre temps*. Montréal: Presses de l'Université de Montréal et Institut du Nouveau Monde.

———. 2016. "Les boutiques des sciences et des savoirs, au croisement entre université et développement local durable." In *Justice cognitive, libre accès et savoirs locaux. Pour une science ouverte justice, au service du développement local durable.* Québec: Éditions science et bien commun. https://scienceetbiencommun.pressbooks.pub/justicecognitive1/chapter/les-boutiques-des-sciences-et-des-savoirs-au-croisement-entre-universite-et-developpement-local-durable/.

Piron, Florence, A. B. Diouf, M. S. D. Madiba, T. H. Mboa Nkoudou, Z. A. Ouangré, D. R. Tessy, H. R. Achaffert, A. Pierre et Z. Lire. 2017. "Le libre accès vu d'Afrique francophone subsaharienne." *Revue française des sciences de l'information et de la communication* 11. https://doi.org/10.4000/rfsic.3292.

Piron, Florence, and Pierre Lasou. 2014. *Pratiques de publications, dépôt institutionnel et perception du libre accès. Enquête auprès des chercheuses et chercheurs de l'Université Laval.* Québec: Université Laval. http://www.bibl.ulaval.ca/fichiers_site/services/libre_acces/pratiques-de-publication-libre-acces.pdf.

Piron, Florence, et al. 2016. "Faire du libre accès un outil de justice cognitive et d'empowerment des universitaires des pays des Suds." *Libre accès aux publications scientifiques entre usage et préservation de la mémoire numérique.* Tunis: CCSD.

Polanco, Xavier. 1990. *Naissance et développement de la science-monde.* Paris: Éditions la découverte/Conseil de l'Europe/UNESCO.

Quijano, Aníbal. 2000. "Coloniality of Power and Eurocentrism in Latin America." *International Sociology* 15 (2): 215–32. https://doi.org/10.1177/0268580900015002005.

Sarr, Felwine. 2016. *Afrotopia.* Paris: Philippe Rey.

Sylla, Cheikh Sidou. 2016. "France: le Service de Gestion des Etudiants Sénégalais à l'Etranger (SGEE) mobilise une importante mine de savoirs au bénéfice de la communauté universitaire." *Diaporas.fr.* October 4, 2016. http://www.diasporas.fr/france-le-sgee-mobilise-un-importante-mine-de-savoirs-au-benefice-de-la-communaute-universitaire/.

Thiong'o, Ngũgĩ wa. 2011. *Décoloniser l'esprit.* Paris: La fabrique éditions.

Vessuri, Hebe. 1994. "The Institutionalization of Western Science in Developing Countries." *The Uncertain Quest. Science Technology and Development.* New York, NY: UNU Press.

Visvanathan, Shiv. 2009. "The Search for Cognitive Justice." *India Seminar.* http://www.india-seminar.com/2009/597/597_shiv_visvanathan.htm.

Wallerstein, Immanuel. 1996. "Restructuration capitaliste et le système-monde." *Agone,* 16: 207–33.

Ziai, Aram. 2007. *Exploring Post-Development: Theory and Practice, Problems and Perspectives.* London: Routledge.

About the Contributors

Serine Haidar Ahmad is a general management consultant at Booz Allen Hamilton, Beirut, Lebanon. She works with clients across different fields ranging from public sector to energy and environment, cybersecurity, and defence.

Sarita Albagli is a senior researcher at the Brazilian Institute of Information in Science and Technology (IBICT) and Professor in the post-graduate program in information science, developed by IBICT and the Federal University of Rio de Janeiro (UFRJ). Since 2004, she has coordinated the Interdisciplinary Laboratory on Information and Knowledge Studies (Liinc) and is editor of *Liinc em Revista*, a scientific open access journal.

Denisse Albornoz, formerly a research associate at OCSDNet from 2015 to 2018, currently works for Hiperderecho, a digital rights organization in Peru, where she investigates gender and technology, data justice, and feminist digital cultures. Her research and advocacy work in India, Canada, and Peru has looked at power and inequality in knowledge production, science, technology, and education.

Irene Agrivine is an open systems advocate, technologist, artist, and educator. A founding member and current director of HONF, the Yogyakarta-based new media and technology laboratory, she participates in numerous festivals and exhibits her work and lectures at galleries, museums, and universities around the world. She is a winner of the 2015 Prix Ars Electronica award, a prestigious European Commission-supported competition for cyberarts in Linz, Austria.

Mahmoud Al-Hindi is Associate Professor in the Department of Chemical and Petroleum Engineering at the American University of Beirut. His research interests include renewable energy desalination, water reuse, membrane processes, removal of emerging contaminants from water, citizen science, and process integration and optimization.

Valeria Arza is a senior researcher in science, technology, and innovation policy at the National Scientific and Technical Research Council and an Associate Professor at the Buenos Aires University. Her areas of expertise and interest include open and collaborative developments, innovation and sustainability, and science, technology and innovation policy.

Victor Odhiambo Awino is pursuing a Master's degree in Research and Public Policy at the University of Nairobi. He is trained in statistical packages and has analytical skills and experience in the statistical analysis of both qualitative and quantitative data, and institutional analysis.

Rima Baalbaki is a doctoral candidate in atmospheric sciences at the Institute for Atmospheric and Earth System Research, University of Helsinki. Her research interests lie broadly within the field of environmental sciences measuring toxicants in air, smoke, and water. As part of the OCSD network, she was project manager of a community-based research campaign to mitigate local water pollution challenges.

Viviana Benavides is an industrial engineer and works as an Assistant Professor at the Universidad Tecnológica de Pereira in Colombia. She is active in several research groups that aim to develop innovative teaching methods using interactive games and tools such as design thinking, and has contributed to initiatives to improve the environmental governance of the Risaralda Model Forest.

Maurice Ochieng Bolo is founding Director of The Scinnovent Centre, a science, technology, and innovation policy think-tank based in Nairobi, Kenya. He has published widely on issues related to innovation, technology, policy, and entrepreneurship, and is a Visiting Research Fellow in the Development Policy and Practice of the Open University (UK) and a Research Associate of the Innogen Institute at the University of Edinburgh, Scotland.

Dora Ann Lange Canhos is Director of Centro de Referência em Informação Ambiental in Brazil, where she works with online biodiversity information systems, focusing on data usefulness and usability, data quality, and indicators. She is a member of the development team of the speciesLink network and of the coordinating team of the National Institute of Science and Technology, Virtual Herbarium of Plants and Fungi.

Vanderlei Perez Canhos is Director of Centro de Referência em Informação Ambiental, where he focuses on strategies and guidelines for the development of public policies aimed at the management and maintenance of information systems on biodiversity.

Leslie Chan is Associate Professor and Associate Director of the Centre for Critical Development Studies at the University of Toronto –Scarborough. He is also Director of Bioline International, an Open Access platform for peer-reviewed journals from Global South countries, and serves on the advisory board of the *Directory of Open Access Journal* and the San Francisco Declaration on Research Assessment.

Sidnei de Souza is Director of Informatics at the Centro de Referência em Informação Ambiental in Brazil, where he is responsible for the design, development, and maintenance of several online information systems, including the speciesLink network, INCT-Virtual Herbarium of Flora and Fungi, Information System of Microbial Collections, Moure's Bee Catalogue, Flora brasiliensis online, and Bioline International.

Hugo Ferpozzi is a postdoctoral fellow at the National Scientific and Technical Research Council, Argentina and a teaching assistant at the University of Buenos Aires, the University of Quilmes, and Maimonides University. His research interests lie broadly within the field of science, technology, and society studies, with a focus on how knowledge production in life sciences shape social and political problems in endemic and non-endemic regions.

Felipe Fonseca is a researcher, curator, writer, teacher, and producer living in Ubatuba, São Paulo, Brazil. He articulates projects that promote collaborative and critical appropriation in the fields of digital technologies, media laboratories, socially-oriented innovation, experimental digital culture, and public policies for science, technology, and innovation.

Laura Foster is Associate Professor of Gender Studies at Indiana University–Bloomington, where she is also an affiliate faculty member of the African Studies Department and Maurer School of Law. Her research broadly focuses on the co-constituted relationships of law and science, and how such interactions historically structure and reinforce certain bodies, identities, knowledges, and practices over others.

Mariano Fressoli is a research adjunct at the Consejo Nacional de Investigaciones Científicas y Técnicas, Argentina, where his research interests include grassroots innovation movements and open and collaborative science. He is co-author of *Grassroots Innovation Movements* and co-organizer of the Congress of Open and Citizen Science in Argentina.

Rebecca Hillyer was the Monitoring and Evaluation Coordinator with OCSDNet from 2015 to 2018. She has worked for community-based research and advocacy organizations across Canada, Ghana, South Africa, and the U.K., where her focus was on using participatory models to promote locally driven change. Since 2017, Rebecca has focused her attention toward understanding how individuals and communities can shape urban life, through her work with municipalities in South Africa and in her hometown of Owen Sound, Canada.

Hermes Huang is a design thinking facilitator based in Bangkok, Thailand, where he focuses on the intersection of the open design process, low-cost technology, and community development. He is a managing partner at InsightPact, a consultancy working to build insightful leaders and inclusive products and services with design thinking, and a lecturer at the Thammasat University School of Global Studies.

Aliya Ibraimova is a project coordinator at the NGO PF CAMP Alatoo in Kyrgyzstan, working in the area of sustainable natural resources management to improve the livelihood of mountain communities. Since 2015, she has been promoting and developing an environmental education and citizen science approach as an innovative tool for improving environmental decision-making in remote mountain communities.

Sammy Kayed is Development Manager at the American University of Beirut, Nature Conservation Center. His research interests include action-oriented citizen science, data, and IT tools for development and participatory approaches to community-driven action on local environmental issues. He has managed the development and launch of fourteen projects and three start-ups focused on co-creating science and living solutions to environmental crises, and has also worked with the US Environmental Protection Agency on marine debris and the US Bureau of Indian Affairs on Native American water rights.

Wassim Kays, an independent mosaics artist in Beirut, joined the Nature Conservation Center in June 2009 as a community outreach coordinator, responsible for all communication between the centre and government ministries, municipalities, and schools. With Salma Talhouk, he developed and tested a manual for extracurricular nature-related activities, which was distributed to all schools in Lebanon.

Denisa Kera is a philosopher and designer who uses prototypes and various creative strategies to involve the public in emerging science and technology issues. She is a Marie Curie research fellow at the University of Salamanca, BISITE group, where she works on block-chain, design, and policy issues. Prior to that, she was an assistant professor at the National University of Singapore, visiting assistant professor at Arizona State University's Center for the Study of Futures, and senior fellow at Prague College.

Pablo Kreimer is a sociologist, principal investigator, and Full Professor of the Sociology of Science and Director of the STS Center in Buenos Aires. He has been working for thirty years on several research projects related to the production and use of knowledge, the relationships between central and peripheral regions, and the co-production of social problems and scientific objects, all of them focused on Latin America.

Josique Lorenzo Lemire was formerly at the Tropical Agricultural Research and Higher Education Centre in Costa Rica, working for the Ibero-American Model Forest Network and as a consultant for different programs at CATIE. She has a background in anthropology and international development, and is highly adaptable, having worked on three continents with a range of stakeholders, from bureaucrats and researchers to farmers and small entrepreneurs.

Maria Lucia Maciel is a retired professor at the Federal University of Rio de Janeiro, coordinator of the Interdisciplinary Laboratory on Information and Knowledge (Liinc) and editor of the journal *Liinc em Revista*. She has worked mainly in the following subjects: science and technology for development; knowledge and innovation; production and circulation of science, scientific dissemination, open science and open innovation.

Leonor Costa Maia is Professor at the Federal University of Pernambuco, curator of the URM Herbarium, and Head of the Mycorrhiza Laboratory, where she coordinates the INCT-Virtual Herbarium of Flora and Fungi. Her primary experience and focus is in the areas of botany and mycology, where she works mainly on collections (herbarium) and taxonomy, as well as the application of fungi, primarily arbuscular mycorrhizal.

Maurice McNaughton is Director of the Centre of Excellence for IT-enabled Innovation at the Mona School of Business and Management, The University of the West Indies. His research focus spans the domain of emerging Open ICT ecosystems, including Open Source software, big/open data, and mobile and cloud computing. He is a founding member of the Caribbean Open Institute, a regional coalition of Caribbean organizations that works with regional governments, researchers, journalists, technologists, NGOs, and academics to raise awareness, strengthen capacity, and foster collaborations toward the adoption of open development approaches.

Cameron Neylon is interested in culture and communities and how they interact with technology, with a specific focus on researchers and research. He has a background in Open Science, including the development of Open Lab Notebooks and Open Notebook Science. His main focus is in developing understanding that will help us to design ecosystems that support open practice in its broadest sense and help us build knowledge systems that empower people.

Thomas Hervé Mboa Nkoudou is a PhD student in Public Communication at Université Laval (Quebec). His involvement as a co-researcher in the SOHA Project (Open Science in Haiti and French-speaking Africa) has given him strong experience in the field of scholarly communication. He is working on improving the visibility of African researchers on the scientific web, promoting diversity and inclusion in Open Access, and confronting the neo-colonial and neo-capitalist hidden faces of Open Access.

Dorine Odongo is a science communicator with more than ten years of experience in developing and implementing research and development uptake strategies. Passionate about research uptake, she serves on the board of the Kenya Network for Dissemination of Agricultural Technologies, where she advises on applied research and livelihood support training.

Angela Okune is a doctoral candidate in the anthropology department at the University of California – Irvine, where she studies data cultures and infrastructures of qualitative research groups working in Kenya in the broader context of equity, knowledge production, and socio-economic development in Africa. From 2010–2015, she was co-founder of the research department at iHub, Nairobi's innovation hub for the tech community. She has been involved with OCSDNet as a Network Coordinator since its inception in 2014.

Aikena Orolbaeva is in the Master's program in landscape ecology and nature conservation at Greifswald University in Germany. She has been working on environmental education for the Kyrgyz Mountains Environmental Education and Citizen Science project and for the Ecosystem-based Adaptation to Climate Change project in Bishkek, Kyrgyzstan, where she was involved in community mobilization issues and agrotourism and extracurricular development activities.

Henrique Zoqui Martins Parra is Associate Professor at the Federal University of São Paulo (Unifesp), in the Social Sciences Department and Post-Graduate Programme in Social Sciences. He is a coordinator of Pimentalab, the laboratory of technology, politics, and knowledge, a research member of the Latin American Network of Surveillance, Technology and Society Studies, and a member of Liinc, the interdisciplinary laboratory on information and knowledge studies.

Florence Piron is an anthropologist, ethicist, Full Professor in the Department of Information and Communication at Laval University, and founding President of the Association for Science and Common Good and its Open Access publishing house, Éditions science et bien commun. She is interested in the links between science, society, and culture, both as a researcher and activist for a science that is more open, inclusive, socially responsible, and focused on the common good. She is involved in several action-research projects aimed at promoting and disseminating in Open Access scholarly knowledge from French-speaking sub-Saharan Africa and Haiti, and at supporting endogenous research capacity in these regions of the world.

Alejandro Posada is a research analyst for Econometria Consultores, a policy evaluation firm based in Colombia, where he is researching the impact of agricultural insurance on farmer risk-management in Guatemala. He was a research associate with OCSDNet and project director with Knowledge G.A.P, based in Colombia, and has international work experience with NGOs, international cooperation agencies, and research networks involved in rural and agrarian development, agricultural financial markets, knowledge production, and Open Access.

Lila Rao-Graham is Deputy Executive Director and a senior lecturer in information systems at the Mona School of Business and Management, The University of the West Indies. Her research interests include business intelligence, data quality, and decision support systems with a particular focus on providing organizations with solutions in order to leverage the value from data for effective decision-making.

Apiwat Ratanawaraha is Associate Professor in the Department of Urban and Regional Planning, Chulalongkorn University, Bangkok. Recent research includes the future of urban life in Thailand, the future of clubs and commons, globalization of land, mapping urban noise levels in Bangkok using citizen science, and informal mobility in Thailand. He has been a Visiting Assistant Professor at MIT's Department of Urban Studies and Planning, and a Visiting Scholar at the Harvard-Yenching Institute.

John Mario Rodriguez is in a PhD program at the University of Girona, Spain. He has been a Professor, Faculty of Environmental Sciences, at the Universidad Tecnológica de Pereira for two decades and, since 2013, at the Faculty of Industrial Engineering. He is

interested in the use of citizen science as a tool to achieve sustainability and citizen participation, and adheres to the Open Science
principles in his day-to-day work.

Aline Rosset is a researcher, teacher, and project manager who has
worked as a Research Fellow at the Mountain Societies Research
Institute, University of Central Asia in Bishkek, Kyrgyzstan. Her main
interests lie in sustainable development and how to combine learning,
environmental research, participatory creation, and dissemination of
local knowledge to promote more equitable, informed, and sustainable
management of natural resources.

Najat A. Saliba is Professor in the Chemistry Department, Faculty of
Arts and Science, American University of Beirut, and Director of the
AUB-Nature Conservation Center. At AUB-NCC, she offers innovative
solutions to local rural environmental challenges through public
participatory and citizen science approaches. She is the recipient of
many awards, including the 2019 L'Oreal-UNESCO International
Award for Women in Science, the National Order of the Cedar from
the President of the Lebanese Republic, and the 2018 American Psychological Association Prize for Interdisciplinary Team Research.

Tobias Schonwetter is Director of the Intellectual Property Unit and
Associate Professor at the University of Cape Town's law faculty. He
is a principal investigator for various intellectual property-related
research and capacity-building projects, including Open AIR. His
specialty is intellectual property, particularly the relationship between
intellectual property, innovation, and development. He is also an
associate member of the Centre for Law, Technology and Society at
the University of Ottawa.

Tommy Surya graduated in graphic design from the Indonesia Institute
of the Arts, Yogyakarta. He is co-founder of the city's new media art
laboratory, HONF (the house of natural fiber), and a key figure behind
Cellsbutton, the Yogyakarta International Media Art Festival, and
YIVF, the Yogyakarta International Videowork Festival. He is also a
founding member of FabLab Yogyakarta (HONFabLab), Indonesia. He
was awarded the DIAGEO-British Council Social Enterprise Challenge
for Arts, Creative and Tourism Organisation prize in 2015.

Salma N. Talhouk is Professor in landscape design and ecosystem management, Faculty of agricultural and food sciences at the American University of Beirut. Her research interests include the promotion of community stewardship of natural resources through decentralization and citizen participation strategies. She is currently leading a participatory mapping project that encourages local residents to identify and document natural and cultural landmarks in their towns. The goal is to produce a nature and heritage conservation, locally mapped national archive built as a free digital platform that is compatible with smartphones.

Halla Thorsteinsdóttir is Adjunct Professor at the Institute of Health Policy Management and Evaluation, University of Toronto and at the Program on Science and Technology Development for Society in Cinvestav, Mexico. She is also Director of Small Globe. Her research focuses on health innovation in low- and middle-income countries where she has examined South-South Collaboration in health biotechnology as well as Canada's health biotechnology cooperation with emerging economies.

Cath Traynor was Program Director, Climate Change at Natural Justice–Lawyers for Communities and the Environment. Her research interests include Indigenous knowledge systems, natural resource governance, climate change impacts and solutions, and implications for communities. As a practitioner, she strives to include community representation in research processes and to co-develop approaches and tools that support communities to engage with external stakeholders on issues related to their land, culture, and environment.

Hebe Vessuri, a retired Professor in science, technology, and society studies from the Venezuelan Institute of Scientific Research, is currently associated with the Universidad Nacional Autónoma de Mexico's Research Center for Environmental Geography, as well as the Patagonian Institute for Social Sciences and Humanities of CENPAT/CONICET in Argentina. More recently, she has been working on the internationalization of the social sciences, the democratization of expert knowledge, and novel forms of doing science.

CPSIA information can be obtained
at www.ICGtesting.com
Printed in the USA
JSHW021607280220
4518JS00006B/125